Rethink Economics and Business Models for Sustainability

Gitte Haar

Rethink Economics and Business Models for Sustainability

Sustainable Leadership based on the Nordic Model

Gitte Haar
Center for Circular Economy
Copenhagen, Denmark

ISBN 978-3-031-56655-4   ISBN 978-3-031-56653-0   (eBook)
https://doi.org/10.1007/978-3-031-56653-0

© The Editor(s) (if applicable) and The Author(s), under exclusive license to Springer Nature Switzerland AG 2024

This work is subject to copyright. All rights are solely and exclusively licensed by the Publisher, whether the whole or part of the material is concerned, specifically the rights of translation, reprinting, reuse of illustrations, recitation, broadcasting, reproduction on microfilms or in any other physical way, and transmission or information storage and retrieval, electronic adaptation, computer software, or by similar or dissimilar methodology now known or hereafter developed.

The use of general descriptive names, registered names, trademarks, service marks, etc. in this publication does not imply, even in the absence of a specific statement, that such names are exempt from the relevant protective laws and regulations and therefore free for general use.

The publisher, the authors, and the editors are safe to assume that the advice and information in this book are believed to be true and accurate at the date of publication. Neither the publisher nor the authors or the editors give a warranty, expressed or implied, with respect to the material contained herein or for any errors or omissions that may have been made. The publisher remains neutral with regard to jurisdictional claims in published maps and institutional affiliations.

This Springer imprint is published by the registered company Springer Nature Switzerland AG
The registered company address is: Gewerbestrasse 11, 6330 Cham, Switzerland

If disposing of this product, please recycle the paper.

Preface

For the past 15 years, I have worked together with ambitious business leaders on the green transition of companies. In these years, the world has changed a lot. The need for change in the way we live and run businesses is now clear to most people. Climate change has set in with rising temperatures, rapid melting of the ice, forest fires, and floods. Human spread and overconsumption harm nature to such an extent that we gradually threaten our own habitats and livelihoods. Eight years ago, we were given the Sustainable Development Goals by the UN and companies have begun to adopt them as a communication platform. I still experience that business leaders grope their way forward when it comes to how to get started working strategically with sustainability and the green transition as a change agenda.

Based on the science, solutions, and methods described in my other book on *the Great Transition to a Green and Circular Economy* (2024) it is necessary to discuss how we create economic models, incentives, and leaders to implement the transition to a fair and sustainable world at high speed. We need to reestablish economic trust and trust in society to create a fair and sustainable world. Our societies and prosperity are driven by economic models that do not account for our basic dependency on nature, raw materials, climate, and biodiversity. Nevertheless, we live on a planet and are part of nature. We must understand that humans cannot successfully and sustainably control nature. We are part of nature and need to subordinate ourselves to the natural ecosystems to regenerate a sustainable and livable planet. We need to decouple growth from the continuous destruction of nature and to regenerate the nature, climate, and biodiversity necessary for human existence.

With this book I hope to contribute to the discussions on rethinking economics and business models to create sustainability. I link this to the new legislation on ESG, climate, and sustainability that is evolving globally these years. I try to provide solutions for society, companies, and organizations and build hope.

The Nordics is a place where many leaders look to find solutions. Being from Denmark and having worked internationally for decades I try to give my view on how the Nordics can provide solutions for societies and businesses. Capitalism must be rethought and access to our natural resources must be regulated much more strongly to create a sustainable living and a planet that can provide for 10 billion people. Maybe the Nordics is where to look to find inspiration for the great transition. The Sustainable Development Goals (SDGs) are not only the world's most important plan, but it is also the most important plan for companies to create a sustainable world and prosperity for all. They are used widely in the Nordics and are also an important framework for this book.

I hope to contribute to seeing companies and capitalism in a new light with inspiration from the Nordic society models. Nordic societies have proven themselves economically and humanly strong in a rapidly changing world. Companies are central to the green transition, and the transition is urgent. Action and ambitious strategies are needed to get started. Therefore, I also hope to contribute to the discussion on sustainable leadership and new values in leaders to drive the change, and to provide solutions using ESG data, tools for companies, and the SDGs in a Nordic perspective as a starting point.

I hereby thank my family for their continuous support in climbing the mountain it was to write this book and the book *The Great Transition to a Green and Circular Economy*, also published by Springer this year (2024). Special thanks to Søren Krasilnikoff for being my private editor providing the push and the critics needed. Thanks to all my four children for being the reason and constant inspiration to pursue dramatic change and having the courage to speak up.

Copenhagen, Denmark   Gitte Haar

About This Book

This book gives a broad narrative on why societies and business landscape are changing dramatically over the next few years, and how businesses are the central wheels in the transformation to a fair, green, and circular economy that is necessary for human existence. Another book by the same author describes the great transition to green and circular economy (Haar, The Great Transition to a Green and Circular Economy, 2024).

This book consists of three parts.

Part I – *Rethinking Economics and Rethink Business Models*—describes the need and conditions to drive change in a very uncertain and unstable world. It builds the common understanding of why the transition to a fair, green, and circular economy is important for businesses and for society.

Part I is about rethinking the economic models and the business models in a situation where sustainability and disruption are dominant. It describes the time when purpose hit the businesses and what was causing it. The Nordic corporations seem more eager to meet the transition even though there are still lots of challenges to be met, especially in the transition to a circular economy. ESG have become the new common language for addressing the challenges, but companies need to figure out how to understand and use the ESG in a business context, also in the Nordics.

This part introduces the drivers of disruption and what this means to society and businesses and how to rethink the economic models for society and the business models for companies. It introduces the purpose as a business driver for companies and sustainable leadership presenting the purpose gap between business leaders of today and the future customers and employees. It elaborates on the need for shift in management and leadership styles to be able to meet future demands. This part includes a description of all the new non-financial data that will meet the companies and how to handle the new reporting and ESG legislation. A chapter in this part also introduces a new way of organizing companies for a circular economy, inspired by M. Porters value chain.

Part II – *Presents the Sustainable Development Goals (SDGs)*—as a framework for business, organizations, and societies in a Nordic perspective. Here is a view from and on the Nordics as societies and the Nordic businesses including an introduction to the Nordics to understand some of the characteristics of the societies here.

This part helps leaders but also others to understand the SDGs and provide new ways of accessing businesses in a sustainable context, providing insights on the urgent need for change and how sustainability and new economic solutions will become the largest strategic challenge that companies will face now and in the future. The SDGs are organized in a Strategy House to assist leaders in all levels of society to prioritize the SDGs. This book reviews the SDGs and puts them into perspective of the Nordics for companies and others as a central wheel for driving the changes also on societal level.

Part III – *Describes the Sustainability Journey and Provides Tools and Methods*—for the sustainability journey and assist companies in implementing and communicating the transition to a green and circular economy. It shows how to build a sustainability roadmap for companies for development of their sustainability strategy.

Contents

Part I Rethink Economics and Business Models

1 Introduction.. 3
2 Disruption 23
3 Rethink Economics and Business Models........................ 31
4 New Company Purpose with Sustainable Leadership............. 49
5 New Economics Need New ESG Data.............................. 67
6 Organizing the Company for the Green and Circular Economy.... 77

Part II Sustainable Development Goals for Transition in a Nordic Perspective

7 Introduction to Sustainable Development Goals and the Nordics... 93
8 Sustainable Development Goals in a Company Perspective........ 97
9 Sustainable Development Goals for Government Bodies and Legislators in a Nordic Perspective 125
10 The Sustainable Development Goals (SDGs) in a Nordic Society Perspective 135

Part III The Sustainability Journey

11 Sustainability Is Complex...................................... 155
12 Sustainability Roadmap and SDGs............................... 163
13 Catalog of Tools and Methods 173
14 Epilog: It Is Just a Human Choice 185

About the Author

Gitte Haar is based in Copenhagen (Denmark) and has advised corporations on transforming and preparing their businesses for the green and circular economy for the past 15 years. She is also the author of several books and whitepapers on the green transition and sustainability based on many years of experience within strategic sustainability.

She is member of non-executive boards of companies working strategically with sustainability, ESG, and the transition to a green and circular economy. Haar holds an MBA from Copenhagen Business School and a Master of Science (Biology/Agronomy) from the University of Copenhagen, Denmark. She has previously served as an international management consultant at Arthur Andersen and Deloitte.

List of Figures

Fig. 1.1	Sustainable Development Goals	6
Fig. 1.2	The Anthropocene Age	8
Fig. 1.3	Global distribution of income in the years 1880, 1975, and 2015	12
Fig. 1.4	UN GHG Protocol for companies	16
Fig. 1.5	Top ten global corporation by market capitalization (US$ billions)	19
Fig. 1.6	Development of the world's largest companies (Fortune 500) from 2009 to 2020	20
Fig. 3.1	Planetary boundaries	33
Fig. 3.2	Doughnut Economics	35
Fig. 3.3	Shift in business case	42
Fig. 3.4	The new business case comparing linear and circular business models	44
Fig. 3.5	Global comparison on GHG LCA of combustion engine and electric passenger cars	46
Fig. 4.1	The purpose gap	52
Fig. 4.2	Competence portrait	56
Fig. 5.1	EU sustainability standards	69
Fig. 5.2	Ecosystem for financial and ESG data	72
Fig. 6.1	Resource hierarchy	79
Fig. 6.2	The circular value chain	80
Fig. 6.3	M. Porter's value chain	81
Fig. 6.4	The circular organization model	82
Fig. 7.1	Sustainable Development Goals (SDGs)	94
Fig. 7.2	Map of the Nordics	95
Fig. 8.1	SDGs are ESG	98
Fig. 8.2	Strategy house	100
Fig. 8.3	Gender equality in management	113
Fig. 9.1	The SDG strategy house	127
Fig. 10.1	GINI index of income in 2015 versus 1990	136

Fig. 10.2	**The basics of the Nordic society model**	138
Fig. 11.1	**Global GHG emission under different scenarios and the emissions gap in 2030 and 2035.**	156
Fig. 11.2	**From harmful to regenerative.**	158
Fig. 11.3	**Global income deciles and associated lifestyle consumption emissions**	161
Fig. 12.1	**Sustainability roadmap**	167
Fig. 12.2	**Implementing sustainability strategy**	170
Fig. 13.1	**Linear value chain**	174
Fig. 13.2	**Example of circular value chains for fiber materials**	174
Fig. 13.3	**Example of circular value chains for construction industry**	175
Fig. 13.4	**Circular value chain.**	176
Fig. 13.5	**UN GHG Protocol for companies**	177
Fig. 13.6	**Materiality assessment.**	178
Fig. 13.7	**SDG strategy house**	179
Fig. 13.8	**EU ESG sustainability standards**	180
Fig. 13.9	**Minimizing climate impact in scope 1 + 2**	181
Fig. 13.10	**Transforming into a circular business model (scope 3)**	182
Fig. 13.11	**Circular organization model.**	183

List of Table

Table 7.1 **The Nordics. The GHG emission stated is scope 1+2 only** 96

Part I
Rethink Economics and Business Models

Part I—*Rethinking Economics and Rethink Business Models*—describe the need and conditions to drive change in a very uncertain and unstable world. It builds the common understanding of why the transition to a fair, green, and circular economy is important for businesses and for society.

Part I is about rethinking the economic models and the business models in a situation where sustainability and disruption are dominant. It describes the time when purpose hit the businesses and what was causing it. The Nordic corporations seem more eager to meet the transition even though there are still lots of challenges to be met, especially in the transition to a circular economy. ESG have become the new common language for addressing the challenges, but companies need to figure out how to understand and use the ESG in a business context, also in the Nordics.

This part introduces the drivers of disruption and what this means to society and businesses and how to rethink the economic models for society and the business models for companies. It introduces the purpose as a business driver for companies and sustainable leadership presenting the purpose gap between business leaders of today and the future customers and employees. It elaborates on the need for shift in management and leadership styles to be able to meet future demands. This part includes a description of all the new non-financial data that will meet the companies and how to handle the new reporting and ESG legislation. A chapter in this part also introduces a new way of organizing companies for a circular economy, inspired by M. Porters value chain.

Chapter 1
Introduction

The need for a systemic change, sustainability, and a great transition to a Green and Circular Economy is urgent, but what will drive the change towards a sustainable and fair planet?

The world has not been this instable since the Cold War at the end of the 1980s. The Cold War and the breakdown of the Soviet Union were followed by a long period of global political and economic stability and growth. That is now ending for various reasons, some mentioned above, especially the entrance of Russia on the war scene and the entrance of China on the political and economic global arena challenging the global economies post-corona. The poor state of planet and climate change are becoming a treat to human lives on all continents. The future generations are facing instability to an extent that has not been known since the years between the World Wars.

The global challenges and instability that ask for change are:
- Poor state of the planet that is threatening human habitats and human living due to climate change and mass distinction of biodiversity and wild nature
- Leak of raw materials from the linear economy and the production of waste resulting in scarcity of most raw materials causing pollution and lost values
- Global value chains creating economic dependence across regions, as well as dependence on minerals and raw materials from political unstable regions
- Financial instability with inflation and increasing interest rates
- Global political instability with increasing war actions on most continents. War in Europe with global involvement resulting in the cut of trade of food, energy, and minerals between EU and Russia

> - Increasing political instability in Asia and internally in the USA
> - Shift of financial power from the USA towards China and Asia, together with a weakened and ununited Europe
> - Polarization of wealth slowing down the spread of prosperity among people, globally

Many are looking for new solutions and new ways out of this situation. Here the Nordic societies can be of inspiration and has reached attention globally as examples of democratic and fair societies with resilient economies.

The global economy has over the last 20 years' time become much more interrelated and the regional economies less independent. Now the Chinese economy is also based on market power and consumption as known from the capitalist economies or Keynesian economic models. The financial crisis in the mid-2000s made the US economy dependent on the Chinese financial markets, and trade between the USA, China, and Europe is now very connected. Especially China has become the global manufacturer and provider of goods for the US and European marketeers. This means that inflation, fluctuating interest rates, and instable financial markets no longer can be isolated regionally.

The global economy is now in a situation of global resource scarcity and linear economies, especially in the old, industrialized countries. The global economy is proven only 8–10% circular (Initiative, Circularity Gap Reporting, 2022) and with even lower circularity rates in the strong consuming countries as, for example, the Nordic countries. This means that over 90% of all virgin natural resources extracted are wasted causing loss of values and environmental disasters.

As the financial power is shifting, the political power is also shifting towards the East. The access to material resources has become a battlefield. The economic conditions are changing due to a planet in need with increasing demands for virgin resources. The economies and businesses need new material loops based on reuse and recycling of raw materials to support economic development and wealth and to become independent on virgin raw materials. The time has come to measure impacts on more parameters than just money.

We need to measure and navigate based on the parameters that human existence dependent upon rather than just based on financial impacts. Human existance depend upon: access to raw materials, a livable planet, and long-term values as stable climate, clean water, wild nature, and biodiversity. A sustainable planet and natural ecosystems create the long-term conditions for food production, development of human health, and resilience against changing weather, natural disasters, and epidemics.

The situation calls for dramatic change and a new global order and new ways of measuring progress is needed. Here the existing economic models come short, and

> Money is not a resource; money is a short-term measure and a man-defined currency to exchange values in a situation of resource abundancy.

especially the lack of time as a parameter in the models is a problem. The economic models do not include the long-term impacts and consequences of short-term financial profits.

Economic measures alone have proven unfit to ensure a sustainable and fair growth. The harvest of virgin resource as, for example, coal is now creating challenges 100 years after the extraction started. These climate impacts are not encountered in the models that measure financial and economic progress and success. We need to account and pay for the externalities dragged on by the creation of short-term economic growth. We need to transform society and to transform business and market conditions. The economy and economic models are the drivers of development of society, businesses, and markets, so this is where to start. We need to include the planetary and human boundaries as parameters in the economic models.

Most decisions are based on financial analyses and economic models, and not on the conditions of the planet necessary for human existence. Economic models promote a hit-and-run strategy and do not facilitate long-term responsible decisions—not on a global level, not on national levels, nor on corporate levels. The consequences of climate change and lack of sustainability foster a need for shift in the economic systems. Legislation on sustainability and ESG (environment, social, governance) is emerging, globally to support a fair and sustainable planet for all. ESG disclosure regulation is the first step to start monitoring businesses and economies on other parameters than just financial returns. This type of non-financial data (ESG) will become more important than financial data and economic models in the future, to manage both businesses and nations. Even so, the changes needed are not solved on national levels or regional levels as the impacts and challenges are occurring at a global level.

Not only does the poor state of the planet and climate change force change, online and social media dominance with uncontrolled, rapid communication among people around the globe also influences political movements in new ways that push the shift of political power and the need for a transition.

Sustainable Development Goals

The UN Sustainable Development Goals (SDGs) visualized in Fig. 1.1 were adopted by world leaders on September 25, 2015, in force as of January 1, 2016 (UN, 2024). The SDGs commit all 196 UN member states to work with ambitious, global, and sustainable development. The SDGs are an unprecedentedly ambitious and transformative development agenda and must be achieved by 2030. In the UN, there was broad agreement that development has not taken place at the desired scale and with the desired pace, despite a goal for global sustainable development adopted in the Brundtland Report in 1986:

"Sustainable development is a development that meets the needs of present generations without compromising the ability of future generations to meet their needs." Brundtland report, UN, 1987

Fig. 1.1 Sustainable Development Goals

Over the past 35 years, since this global definition of sustainability came out, we have not reached a sustainable development—on the contrary. Therefore, the SDGs are the new plan for the transition to a sustainable and fair world for people and planet, and they are especially finding ground in educational institutions and municipalities in the Nordics, but not globally and as broadly as could be wished for.

A new business humanism is emerging, and companies are starting to use the SDGs actively in their communication. Especially in the Nordic region, the SDGs have found ground and are being used by companies to frame and substantiate sustainable communication. Often ESG and sustainability is still used to communicate what companies already do, and less the backbone for driving change. There is a lack of strategic implementation of sustainability in companies to meet a greater purpose. Additionally, many companies are fumbling in their work with the new disclosure requirements, ESG reporting, and implementation of the challenges addressed by the SDGs.

When engaging in the SDGs, it is important to read the 169 targets carefully and the facts behind. This is an incredibly educational process. Unfortunately, many throw themselves directly into choosing from the 17 goals at a very general level, but it is in the targets that the challenges and alignments for companies are found.

Eight years after the SDG launch, we are in a hurry to reach these 2030 goals. World leaders are not taking the challenges addressed seriously and are not implementing the visions and legislative frameworks for societies to transform. This is also becoming very clear from the outcomes of the COP28 in the UAE this year (2023). Now the threats from climate change, poor ecosystems, and political and financial instability are becoming urgent, the change is necessary, and politicians are still reluctant to actual commitments.

Companies must take a new stake in this global situation. The unlimited growth in the old, industrialized countries since World War 2 and the globalization of trade since 1980s are coming to an end. Humanity is now limited by the planetary boundaries that have now been exceeded from centuries of human activity. Changing climate is minimizing the zones suitable for human life and food production. Political instability and resource scarcity are limiting access to raw materials, especially of the old, industrialized countries and the alliance of the Western world, causing financial instability. Now it seems as if the financial powers are anti-globalizing. The time were a company can sell and produce anything as cheap as possible is over. Now, we see the long-term consequences of overconsumption based on financial wins and planetary downs.

Companies and trade are the central wheels of driving the changes needed, and the market conditions set by politicians on regional levels are defining how rapid this great transition will happen. Therefore, companies must be involved in new ways and extend their responsibility for a fair and sustainable planet.

The global financial instability is an aggregation of other instabilities. When only the economic outputs are measured, and no accountability is in place for the exploitation of externalities, we will continue to view this as independent poly-crises. Many talks about a new situation of poly-crisis for businesses. Maybe it is many crises deriving from the same fundamental problem, namely, humans spread and lack sustainable and fair living, globally throughout decades.

Impacts on the planet from human activity are difficult to measure, and more importantly political leaders and business leaders will not be able to manage the necessary change and create fair and sustainable lives until we are able to measure these impacts. Lots of new politicians and businesses arise in the dawn of the new planetary and global situation. The call for urgent change is now loud as climate change, and breakdown of ecosystems is coming at high speed. The polarization of economy and political power is also changing rapidly creating completely new market conditions, new citizen requirements, and new customer behavior.

The changes needed in a business and society perspective are also called a modernization of the industrialization, facilitated also by the Internet, AI, and Industry 4.0. We need to transform society, existing corporations, and the economic systems.

Frameworks to support change are the SDGs and the ESGs. The EU Green Deal, as the roadmap for changing Europe, is based on the SDGs, and together with the new legislation coming out in the USA on climate and environmental justice, ESG, and the Inflation Regulation Act (IRA), the change is slowly arising from the political levels. China is rapidly shifting to a Green and Circular Economy and is in the forefront of the energy transition, not only installing most of the global renewable energy (RE) but also producing it. China has a high consumption of coal for energy production (approx. 50% of all coal globally) and manufactures many of the linear products with no potential for recycling. Still, the speed of the Chinese transition is impressive, and their understanding of the geopolitical situation and the need for securing access to raw materials seems much more mature than in North America and the EU. The old Western regions are mostly challenged by domestic political issues and the invasion of Ukraine by Russia and now also the War in the Gaza Strib. Globally the political balances are affected by instability and increasing conflicts in the Middle East and Africa, also linked to abolishing the fossil economy and raw material scarcity in general, as Africa and the Middle East sits on many of the rare minerals that we are dependent upon.

Companies should engage in the SDGs to find the solution on a global scale and on regional levels. The SDGs are not written for companies, but they still offer insight into magnitudes of challenges. Later in this book, tools are provided to help companies prioritize and use the SDGs as strategic drives.

> Sustainability is no longer something to do on the side of traditional business; it is a new business imperative that will change companies and markets.

The Anthropocene Age

The state of the planet that is causing the poly-crises is illustrated in the Anthropocene Age in Fig. 1.2 (Haar, The Great Transition to a Green and Circular Economy, 2024c).

The Anthropocene Age is a new term that covers the fact that the entire planet and all ecosystems are now subject to human dominance. Humans have access to every place on earth and control all ecosystems and organisms. At the same time, we

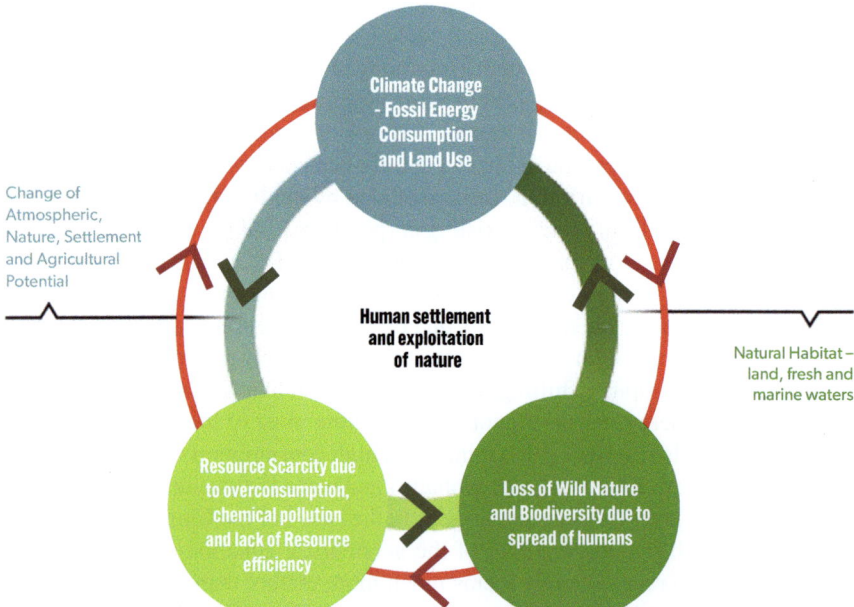

Fig. 1.2 The Anthropocene Age. Human dominance has caused three major crises—climate change, breakdown of ecosystems and biodiversity, and resource scarcity. Now these crises are affecting each other negatively (red circle). A shift to sustainable and fair living will redirect this and make them affect each other positively (green circle)

experience environmental disasters, as climate change; plastic pollution (micro and macro) in the natural environments—especially in aquatic environments; and dramatic decline in biodiversity and wild nature. Industrialized food production makes people suffer from obesity, creates intensive monocultures in agriculture and forestry, and causes chemical pollution of fresh water, oceans, and soils.

Chemical pollution of the products that we use daily is also an increasing problem, and we are now experiencing incalculable health consequences of chemicals in food, clothing, furniture, building materials, and other things. The Anthroprocene Age is a result of the overconsumption in the old, industrialized countries. The overconsumption of fossil fuels, food, and products by the 10% richest globally has put the planet and humanity in a fearful future. The state of the planet and understanding the connections between climate impacts, other sustainability impacts, and a holistic transition to sustainable solutions (the Climate Nexus) is described in another book by the same author (Haar, The Great Transition to a Green and Circular Economy, 2024c).

Changing Conditions

There is a dawning tendency for business leaders to communicate new values other than just financial profit and share value. Companies are talking about their extended responsibilities and shared values with the society, employees, and customers in a broader sense. Companies' role in society and their approach to the outside world is changing strongly supported by legislative changes in both EU and North America. Learnings have come out of the dependence and risk of corona and the Russian invasion of Ukraine, and companies are expected to take a stand also in strong political situations.

EU legislation is putting an extended social responsibility and a responsibility of the extended value (supply) chain on corporations these years. New requirements for products, as the sustainable product initiative, eco-design criteria, and extended producer responsibility, all make businesses the wheels of a great transition towards sustainability (Haar, Chap. 6, "EU regulation to a Green Economy, 2024a").

In the USA, the transition is also driven by legislation as the Inflation Regulation Act (IRA) that to a large extend is climate regulation, the new executive order to revitalize Environmental Justice for All (2023), state regulation on ESG investments, and the new and very rapid emerging markets for carbon removal. Carbon removal has become the new commodity in the USA. China is transforming in a centralized manner due to the geopolitical risks of dependency on fossil fuels. Thereby various reasoning, legislation, and market conditions are driving the change.

> **The new Executive Order to revitalize Environmental Justice for All put on all federal agencies in the USA (White House, USA, 2023) stating:**
> "Environmental justice" means the just treatment and meaningful involvement of all people, regardless of income, race, color, national origin, Tribal affiliation, or disability, in agency decision-making and other Federal activities that affect human health and the environment so that people:
> (i) are fully protected from disproportionate and adverse human health and environmental effects (including risks) and hazards, including those related to climate change, the cumulative impacts of environmental and other burdens, and the legacy of racism or other structural or systemic barriers; and
> (ii) have equitable access to a healthy, sustainable, and resilient environment in which to live, play, work, learn, grow, worship, and engage in cultural and subsistence practices."

All this new impact regulation together with various national constitutions and human rights regulation open the possibility for citizens and organizations to press charges on corporations' historic impacts on environment and climate for them to pay their environmental debt. accumulated over decades. Lawsuits have been passed in Holland, in the USA, and will be common in the future for states and corporations to take responsibility for the destruction of nature and climate. The COP28 companies have announced that they will pay their climate debt. Either directly by investing in land and forest to regenerate nature or indirectly by providing finance to the UN climate fund for securing and financing the transition of the Global South. The developing countries suffering severely from climate change and typically countries that have insignificant GHG emissions. As an example, the Danish company Velux established a Climate Fund to donate €135 Mio. over 10 years (2015–2025) and provide €40 Mio. to regenerate forest in the South (Uganda, Madagascar, and Vietnam) as compensation for Velux' historic GHG emissions of 4.5 million tons CO_2 in the last 80 years of operation, compared to an annual profit of €85 Mio. (2022). American corporations as Google and Microsoft have done some in the same direction. Thereby, not only legislation but also corporations are paving the way for companies not only to become climate neutral in the future but also to compensate for historic emissions and thereby contributing to regeneration of nature and climate.

The evolving ESG agenda is also affecting the responsibilities of executives, and this is materializing in board insurances of executives. It is now a reality that some insurance companies will not take out board insurance for executive management in companies without a clear and implemented ESG profile.

Economic and Political Development

The capitalist economy has reached an impasse. Many have for a long time believed that democracy and the free market created a fair distribution of wealth and abolished poverty. This is no longer the truth and the old, Western democracies experience increasing economic inequality and increasing accumulation of capital on very few hands. This creates new challenges and financial instability. The planetary challenges in the Anthropocene Age with climate change, loss of wild nature and biodiversity, and resource scarcity are also contributing to economically and politically instability with increasing inflation and interest rates (Haar, Part I: Introducing the need for a Green and Circular Economy, 2024b).

Nevertheless, over the past 30–40 years, many people have left poverty. The distribution of wealth since World War 2 has increased even since the fall of colonization in the middle of the twentieth century.

Figure 1.3 shows that especially since 1975, many people in Asia have emerged from poverty, and this is partly due to the globalization (Max Roser, 2013). This does not change the fact that economic inequality is increasing these years and that consumption based on global value chains affect our planet and people creating a new political situation and a new business situation.

Capital is accumulating on fewer hands, creating inequality and instability. Today, the 22 richest men own more than all African women, and the world's 2000 richest people own more than 60% of all the assets, globally equal to what 4.6 billion people own (OXFAM, 2020). The growing inequality is an important topic when world leaders meet. Information to illuminate the problems are presented at the World Economic Forum in Davos every year. Especially, after the financial crisis in the 2006–2013, the rich have become richer, and the poor have become poorer. As an example, 2% of Jeff Bezos' fortune (owner of Amazon) could cover the entire health budget of Ethiopia, one of the populous African countries with 120 million Ethiopians. This is not a fair distribution of wealth.

Business leaders are gaining a greater awareness of the importance of the environment and sustainability. A new age of enlightenment or a new formation of business leaders is slowly emerging, and companies must (again) play a central role in society. As opposed to the fact that for a long time, it has been completely accepted just to generate profits and capital gains for the shareholders, now there is an increasing focus on how companies behave and contribute to society; how they treat their employees, customers, suppliers, and society; and especially their impacts on climate and the environment. This shift towards shared values and awareness is necessary after decades of creating longer and longer value chains in a global, linear economy. Companies have increasingly sub-optimized a small part of the value chains and supply chains, and the production has been outsourced to areas far away from the consumers, especially to Asia.

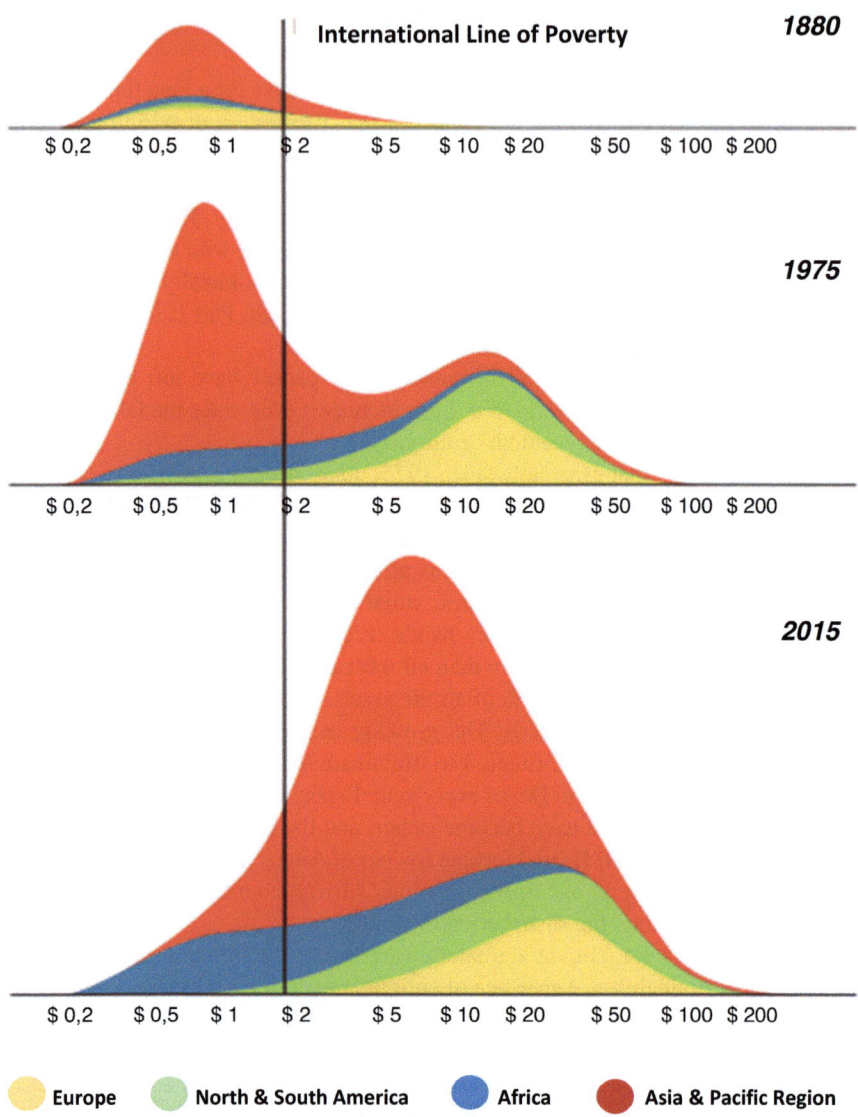

Fig. 1.3 Global distribution of income in the years 1880, 1975, and 2015. Daily income per citizen in US$ 2011—prices showing the economic development. Logarithmic axis. The figures show population growth over the last 125 years as well as the shift to increased wealth and population above the line of poverty

The Stake of Companies in the Great Transition

Greenwashing is a large problem, and some global corporates have been caught engaging in dishonest business and false marketing on their ESG impacts. The systematic fraud conducted by the German car industry on CO_2 emissions from their cars is one of the examples on business irresponsibility. The oil industry has since the 1960s been aware of climate change from fossil energy consumption without acting and without taking responsibility for the damages the industry is imposing on the climate, nature, and people. On the contrary, both the automotive and oil industries have spent billions to obscure these climate impacts. It has been revealed in recent years, especially in the USA, that the oil industry has established think tanks and bought "scientists" to spread falsehoods and cast doubt on independent research on climate change. Financial connections have emerged between even the large, reputable universities in the USA. Many scientists believe that this systematic misrepresentation by industry on climate data is the reason why the recommendations of the Intergovernmental Panel on Climate Change (IPCC, n.d.) warnings have not come out clearly and widely to the public since mid-1980s. The scientific reports of the Intergovernmental Panel on Climate Change (IPCC) in the last three decades have successfully been questioned and miscredited by people, politicians, and business leaders.

Uncertainty and misinformation in disclosing the impacts from the long value chains and products on the environment are also seen in industries other than the oil and automotive industries, such as the sugar industry and the tobacco industry, but also other industries work with hidden knowledge about the impacts of their production on the environment.

In an open society with generally widespread knowledge and data on the negative impacts from businesses, the lack of responsibility is no longer accepted by citizens. Now is the time for companies to take broader responsibility and develop business models and products that do not harm the environment and people. It is a sound liberal principle that the polluter pays for the damage and pollution caused, also being a business. Now we need to live in a way where consumption and businesses do not harm our planet and the people in the value chains. Sustainable living is only in the very rural and developing areas, and economic growth is based in unsustainable societies and businesses because we do not account for the derived effects of consumption on externalities like climate, nature, and human rights.

For many years, companies have based their business on marketing all the products that they can make customers buy. No legislation or accounting for means of production or on the impacts on the planet or people was implemented throughout the full value (supply) chains. This has been supported by politicians' one-sided focus on the GDP growth and state budgets as the way to create wealth. Responsibility for environmental and social impacts has increasingly been the responsibility of the consumers and citizens and not the companies. The time is over where companies consciously or unconsciously can ignore the negative impacts from products and business models on the planet and on the societies in a global perspective.

Many business leaders today do not know the environmental or social footprints of the full value (supply) chain or of their products or services, due to long and untransparent supply chains upstream and downstream. Companies have lost the track to measure or even understand their impacts because the only disclosure requirements were on the financial results. Now, this is changing with extensive legal requirements from states, from institutional investors and stock exchanges on

ESG data. It is rolled out in the EU as of 2021 and ongoing, and now the USA is also implementing disclosures on ESG state-wise.

Today consumers and citizens are not able to understand or identify these complex value chains or evaluate sustainability and separate greenwashing from genuine sustainability. Many consumers and procurers are demanding detailed information. The skilled consumers are to some extent able to extract this information online and from scientific data, but overall consumers and citizens are confused when it comes to making sustainable choices. In the hunt for cheap products, companies in the old, industrialized part of the world have sourced their products from places where they know that both labor conditions are poor and environmental requirements are low. This has resulted in a disconnection of the understanding of how things and stuff is manufactured and the quality of these, causing untransparent and untraceable supply.

Global leaders have become keynote speakers on sustainability, but companies must find their real purpose, and not just a new marketing gig or a new set of communication cards based on SDGs. They need to link close to the products and their full value chains. Then genuine sustainable change can happen, also driven by new consumption patterns.

In these years, many companies experience that sustainable and circular business models are more profitable than the old linear ones. Sustainability is and will increasingly be profitable and a necessary competitive advantage for many companies. Sustainability no longer contradicts profit. Products and companies are rapidly regulated globally against the negative environmental impacts and have for years been regulated on social impacts. Meeting this regulation will facilitate the development of new sustainable solutions and products. Then, sustainability and business go together enhancing each other, instead of either-or.

> Sustainability is a market condition, and data are the key to meet the demands from customers and legislators.

New Business Imperative

New limitation must be set for the liberal (free) markets and on the capitalistic economic model. In the long perspective, creating wealth is much more than financial security. It is necessary to ensure that companies and nations cover the costs of exploiting natural resources, wild ecosystems, people, and other externalities, such as the climate. In fact, the planet is in a state where regeneration is needed rather than just zero impacts as required by most sustainability measures. The commandments are to discuss regulation to create a real free market, new economy models, and the ownership structures together with new sets of values and new types of leaderships for business leaders.

This book is a contribution to the debate on rethinking economic and on how to create a sustainable and fair planet based on sustainable economies and a more equal distribution of wealth.

> **There is a need for rethinking economics and business models that embeds:**
> - A **greater understanding of the actual situation of the planet** and the roadmap to genuine sustainability, a fair planet, and the greater purpose of leaders—political as well as business leaders. Now lack of knowledge is the largest barrier for the great transition.
> - The **will and courage to embrace the changes** and the uncertainties this bring on society, people, and businesses. New leadership skills and understanding the new economics and the genuine sustainable solutions in a holistic way and facilitating the transition with the technologies available as robots, digitalization, AI, and a full online interconnected world.
> - **Full transparency and traceability** are needed to create trust and belief which again requires a huge number of new types of data that are registered, managed, and disclosed as known from the financial data that we are so dependent upon. We only see the tip of the iceberg on the data needed in the future at all levels: society, company, and product level.
> - **Carrying the costs for the use of externalities** and impacts is necessarily possible, and access to all these new non-financial (ESG) data at all levels will make it possible to pay the full price for our consumption.

Only when knowledge has been built, new skills and shared values have been implemented, and data made available on all levels in society, in businesses, and among citizens companies are able to manage and transform towards the new economy.

Unfortunately, the task is not only solved with new business models in companies, new responsible products, and production chains. There is also a need for new types of ownerships of companies. The ownership of assets and shares on capital markets has shown it limitations. Thereby more equal distribution of earned profits and increased asset values is needed. Accounting for the impacts on externalities as climate, planet, and society on a larger scale is also necessary. This is to balance the strong accumulation of capital that threatens the cohesion of especially the old, industrialized countries, such as Europe, the USA, Great Britain, and all over the world.

Another element that is changing the business imperative these years is the UN GHG Protocol for companies defining company impacts in the full value chain including downstream and upstream (scope 3). See Fig. 1.4 on the definitions of the three scopes that has become the basis for all communication and reporting on company impacts in the value chain, globally ((WRI), World Resource Institute and

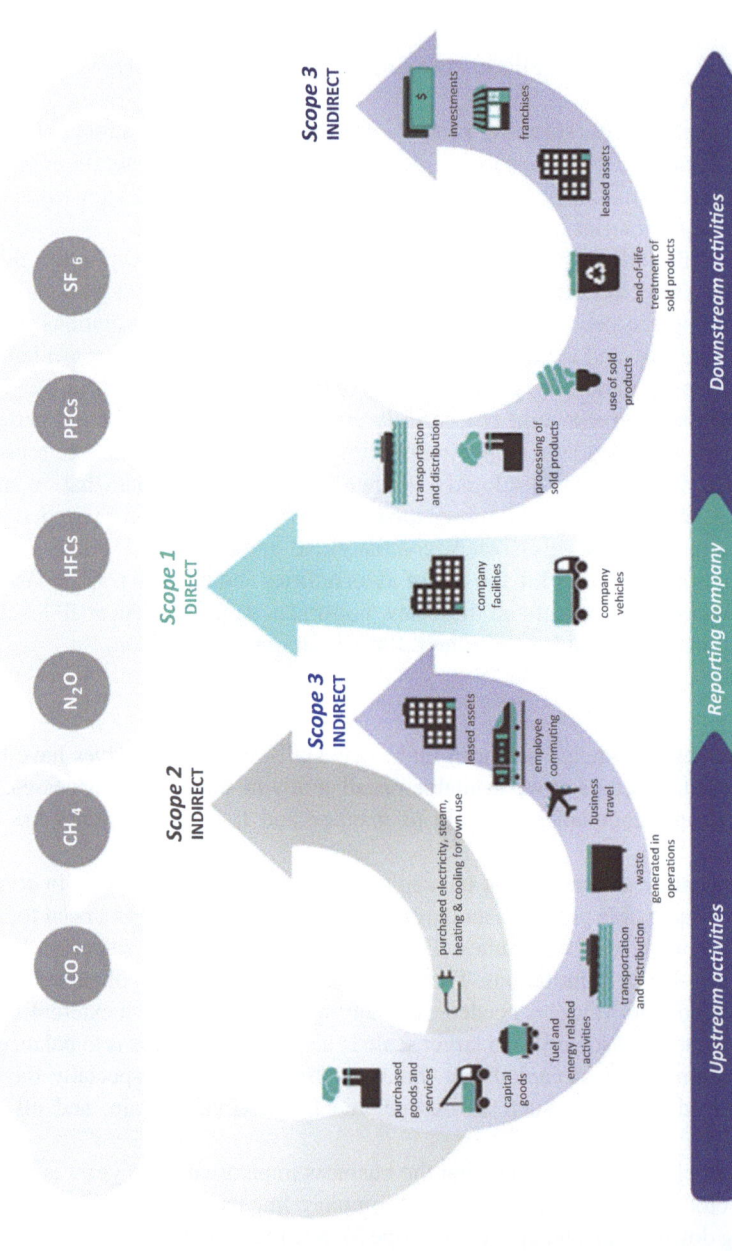

Fig. 1.4 UN GHG Protocol for companies. This protocol delivers a universal understanding of the impacts of companies in the full value chain, divided into three scopes. This is the basis for most regulation on sustainability and ESG, not only on climate impact but all ESG impacts

World Business Council for Sustainable Development (WBCSD), n.d.). The EU has with the Corporate Sustainable Due Diligence (to be passed in 2024) set new standards of fostering sustainability in corporate governance and management systems and thereby widening the responsibility of management to scope 3. Companies that want to put products on the European market in the future must fully account for impacts also in scope 3 and prove that they will minimize their impacts in the full value chain. This will change the way companies source, manufacture, and sell their products and services.

Company Purpose: Sustainability as a Condition

The great transition and the demands for sustainability place new demands on leadership and corporate strategy. Rethinking economics is a paradigm and entails a completely different way of managing businesses. A new economy has a different and more overall long-term goal of creating a sustainable and fair planet.

Many companies have a strong focus on how to utilize digitalization, the Internet, SoMe, AI, and new technologies (Industry 4.0). Industry 4.0 is a large part of a journey towards sustainability and sustainable business models. First business leaders must decide on the direction towards the new market conditions, and not just on how to get there with technology.

> Technology is the *how*, sustainability is the *why*.

All these new technologies and the new access and generation of data is not a goal in itself, but it may facilitate the transition when used right. Many of the IT billionaires of today believed that they revolutionized and democratized the world with the Internet and their new services. They did change the world, but not in the way that they thought because they did not have the overall purpose in place. They thought that the revolution came from the bottom with their innovation and online services for all. It just put them in the front seat of a new and very rapidly developing economy based on free access to the new *raw material* of consumer and citizen data. They forgot the greater purpose of that transition. Business leaders are now forced to set a new direction in a rapid changing world with new sets of shared values and ways of leadership that are much more long term than they are used to.

Businesses are important wheels in the transition to a new economy, if not the most important wheel, but they must find their greater purpose and not just let new technologies drag them into new business opportunities. It seems as if some global political leaders are taking on the responsibility of creating new market conditions and limit free and unpaid access to our planetary resources. But there is still a lack of political will to initiate and implement the great transition in many countries. This makes it even more challenging for business leaders to find the direction in a time of uncertainty and political disorientation. Business leaders must understand the new conditions and requirements to attract future consumers and employees and to create value for owners and society in the long term. The roadmap set by EU Green Deal is good place to look for business leaders on the European markets.

Several international surveys among listed companies and stock exchanges show that companies having a strategic focus on ESG (sustainability, gender equality, and corporate governance) generate better returns. We are facing a shift in leadership with demands for new shared values. Businesses depend on insights into the changing world and a new set of values that the consumers and employees of the future demand from business. Leaders must be able to navigate in a world where larger issues and impacts than company profit and share value matters. Future leadership depends on a strategic understanding of the positive impact that companies have on society and the planet and how the company can contribute. It also requires greater human understanding and societal insight than has been required of leaders in recent decades. Some call it soft skills; actually it is hard work because it requires much more genuine, transparent, and personal leadership. Especially the young employees demand from their leaders to show more of themselves and their personal values to appear credible in a purpose-driven company.

Increased focus on the overall purpose of the company in society and the contribution to changing the world may mean a shift towards more feminine and holistic human values in management. New values are needed to meet the new demands from society, customers, investors, and especially employees. Hopefully this results in a greater diversity in executive management of the companies, including more women in strategic management and on the boards along with diversity of race, sexual orientation, etc. Companies are only able to deliver a complete strategy and execution of values when they mirror the diversity of their customers, employees, and the society they depend upon.

Change of Business Landscape

Companies not preparing for the transition and not daring to make long-term investments in sustainable solutions are challenged. They will not be able to react quickly enough to the changes that the world is bringing. This means that old companies will be disrupted by new players who are not bound by old structures, old capital, old technologies, and old-time investments and long value chains. Thus, the transition will rise from disrupters rather than from the existing global corporations. This trend of disruption is well-known, and the Internet and Industry 4.0 are also causing disruption of market access. The largest corporations today are not the same as 20 years ago in terms of market capitalization. Figure 1.5 shows a shift from the automotive and oil industries to ICT (information and communication

Change of Business Landscape 19

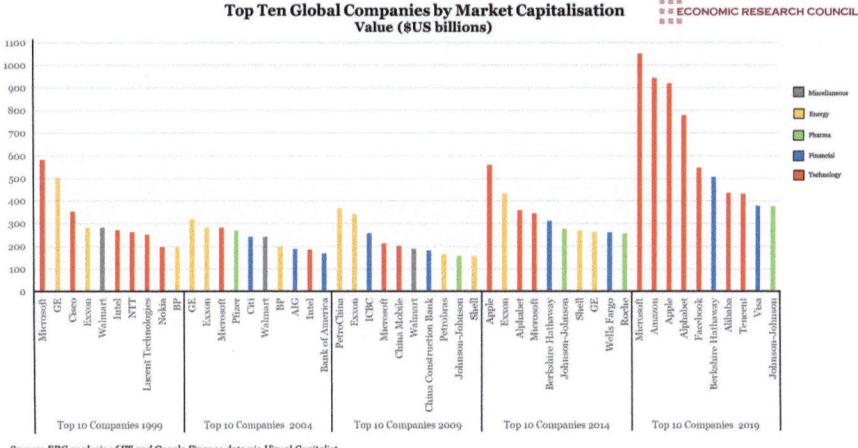

Fig. 1.5 **Top ten global corporation by market capitalization (US$ billions).** Development of dominating corporations and industries over 20 years

technologies), where Internet-based corporations hold the global dominance today (Visual Capitalist, 2019). As of 2023 the ranking is Apple, Microsoft, Saudi Aramco (Saudi Arabia), Alphabet (Google), Amazon, NVIDIA, Tesla, Berkshire Hathaway, Meta Platforms (Facebook), and Eli Lilly. US corporations are still dominating with 9/10 (Economic Research Counsil, 2019).

The transition to a Green and Circular Economy will also change the business landscape of dominating companies and technologies over the next 10–20 years globally, and the renewable energy sector are making its entry on the above list because of this transition. Watch an interesting simulation of the Fortune 5001 over the past 30 years. (https://www.youtube.com/watch?v=fobx4wIS6W0 af WawamuStats).

As mentioned earlier in this chapter, another interesting aspect is the geographic shift in economic power that has occurred since the financial crisis. Figure 1.6 shows the economic development of countries/regions measured by the number of companies and by aggregated turnover on country level from 2009 to 2020 (Qlik, 2020).

Here it is clear that China has become a dominating economy since the financial crisis. The financial crisis hit the USA and Europe hard, and at the same time outsourcing from both the USA and Europe took hold in China. Today, the USA (25,500 US$ billions), the European Union (17,400 US$ billions), and China (18,000 US$ billions) are the largest global economies. Tech companies were to a large extent situated in the USA until pre-corona, and now China is producing a very large share of the technology and clean tech globally, both Microchips, Electrical Vehicles, and Renewable Energy.

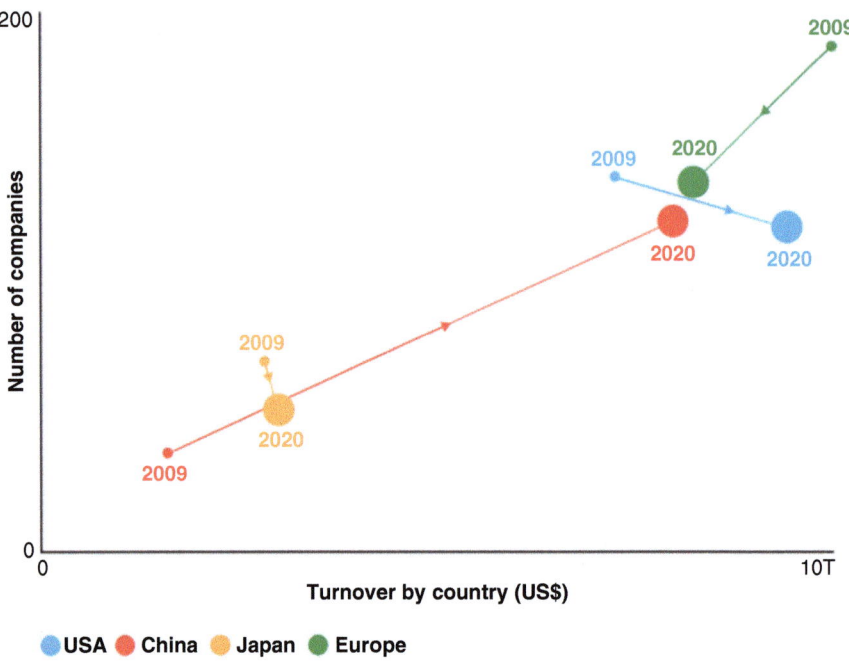

Fig. 1.6 Development of the world's largest companies (Fortune 500) from 2009 to 2020. Total company revenue for the countries from Fortune 500 corporations. Europe: Astria, Belgium, UK Curacao, Denmark, Finland, France, Germany, Hungary, Ireland, Italy, Holland, Polen, Portugal, Russia, Spain, Sweden, Switzerland

The Great Transition in a Nordic Perspective

The Nordics are getting a lot of attention these years from all over the world, as an attraction to the Nordic lifestyle, design, and society model. The Nordic societies have proven resilient and innovative through the global crises since the evolvement of a strong global economy in the last three to four decades. The economies and the democracies of the old, industrialized countries are becoming instable, and the welfare systems are under strong pressure creating a new kind of poverty here. The Nordics seem to have found a way through this that has proven strong. The Nordic countries have developed a society and a business environment that has proven financially strong and resilient and demonstrates a spread of wealth to the populations.

The Nordic economies are closely tied to the EU, especially to Germany and Great Britain as important trade partners, and now also to the USA as a huge marketplace for Nordic companies. The development of the economic success of the Nordics is not based on local economy; it is based on a strong international position.

Nordic corporations are performing well on the global business scene. A study from 2023 showed that the Nordic countries, especially Sweden and Denmark, outperformed on global corporations against country GDP (Corporate Knights, 2023). Nordic companies are also overrepresented on the list of the World's 100 most sustainable companies. Here the Nordics (2021) have 14 companies on the list, compared to the USA with 20 companies. The Nordics have a population at the size of Texas, an aggregated GDP of $1.8 trillion ($66,900 per capita), compared to Texas with $2.4 trillion. The Nordics are the 12th largest economy globally. About 60 Nordic companies qualify on the Forbes list of the World's 2000 largest publicly listed companies. This significantly exceeds the number for Germany, although the combined size of the Nordic economies is less than half of that of that country (Per Lekvall with contributing national author, 2014).

The entrepreneurial atmosphere, the high educational levels, and the old and stable democracies make up an attractive business environment that fosters some of the largest and most sustainable corporations in the world. This also be a reason to look towards the Nordics for a society to copy.

A special management style and business culture exist in the Nordics receiving attention from the outside. The special Nordic business culture is characterized by the lack of trust in authorities in the daily business life, an informal tone, flat hierarchy, and the ability and obligation to speak out against authorities. Holding a title and having structural power does not give the same status and actual power here as in many other parts of the world. In the Nordics, it is more the insights and the ability to behave fairly and decent that fosters the informal power. This spread of power and modesty is in Denmark and Norway described as the Law of Jante, "Do not think highly of yourself," and it takes a while for foreigners, especially Americans, to understand this force. All this is very well described by Robert Strand, a leading expert on Nordic Sustainable Business and Nordic Capitalism at Haas Berkeley University in California (Strand, 2014).

A management book written by an American, Chris Shern, and a Dane, Henrik Jeberg, on the Nordic leadership style tells the story of the rich heritage of the Nordic peoples: as explorers, navigators, trendsetters, and political and educational innovators (Jeberg, 2018).

In a report about the Nordic region published in February 2013, *The Economist* used the title "The next supermodel," pointing to the fact that the Nordic countries cluster at the top of global league tables of everything from economic competitiveness to social health (Leaders, 2013).

Nevertheless, the Nordics are some of the most consuming and waste-generating populations, globally, and the need for a transition to a Green and Circular Economy is larger here than many places.

The overall framework for the companies and markets in the Nordics in the transition to a Green and Circular Economy is the EU Green Deal that was introduced in 2020 including Mobilizing the Industry for a Clean and Circular Economy, Strategy for a Circular Bioeconomy, Farm-to-Fork, and a Biodiversity Strategy. This is framing the transition to a Green and Circular Economy for all sectors and targeted legislation is now imposed for the nations to implement. The EU Green Deal includes these very important pillars for the Nordics and for the whole EU.

The companies that will dominate the marketplace after the transition are those who move first in the direction of a Green and Circular Economy, and this will be the companies that are able to embed the greater purpose; can educate their management, employees, and stakeholders; and are agile and can implement at a speed that will hit the market first—both the consumer market and the employee market. With the impressive statistics on the Nordic companies, stated above, their power and agility towards sustainability strongly supported by social welfare, and free education offered by the Nordic states, it seems as if they will hold an even stronger position in the future, globally.

References

Corporate Knights. (2023). *Global 100 companines driving the transition to a low-carbon, circular economy.* https://www.corporateknights.com/rankings/global-100-rankings/2023-global-100-rankings/2023-global-100-most-sustainable-companies/ Corporate Knights.
Economic Research Counsil. (2019). *Analysis of FT nd Google Finance data via Visual Capitalist.* https://ercouncil.org/2019/top-ten-companies-by-market-cap-over-20-years/.
Haar, G. (2024a). Chapter 6: EU regulation to a green economy. In G. Haar (Ed.), *The Great transition to a green and circular economy.* Springer.
Haar, G. (2024b). Part I: Introducing the need for a green and circular economy. In G. Haar (Ed.), *The Great transition to a green and circular economy.* Springer.
Haar, G. (2024c). *The Great transition to a green and circular economy.* Springer.
Initiative, Circularity Gap Reporting. (2022). *Circularity gap report.* Circle Economy.
(IPCC), I. P. (n.d.). *Climate reports.* https://www.ipcc.ch/reports/: UN.
Jeberg, C. S. (2018). *Return of the vikings – Nordic leadership in the times of extreme change.* Psykologisk Forlag.
Leaders. (2013). The next supermodel. *The Economist,* 02/02.
Max Roser. (2013). *Global economic inequality.* https://ourworldindata.org/global-economic-inequality. OECD dataset.
OXFAM. (2020). *Inequality report presented at world economic forum, Davos.* OXFAM.
Per Lekvall with contributing national author. (2014). *The Nordic corporate governance model.* SNS Förlag.
Qlik, F. 5. (2020). *Global 500.* Retrieved from The history of global 500: https://community.qlik.com/t5/Design/Introducing-the-History-of-the-Fortune-Global-500/ba-p/1740778
Strand, R. (2014). Corporate social responsibility and sustainability in Scandinavia. An overveiw. *Journal of Business Ethics, 127,* 1–15.
Visual Capitalist. (2019). *ERC analyis of FT and Google Finance Data.* Retrieved from ER Counsil: https://ercouncil.org/2019/top-ten-companies-by-market-cap-over-20-years/
White House, USA. (2023, April 23). *Executive order on revitalizing our nations commitment to environmental justice for all.* Precident Biden, USA. https://www.whitehouse.gov/briefing-room/presidential-actions/2023/04/21/executive-order-on-revitalizing-our-nations-commitment-to-environmental-justice-for-all/.
(WRI), World Resource Institute & World Business Counsil for Sustainable Development (WBCSD). (n.d.). *GHG Protocol for companies.* Retrieved from GHG Protocol: https://ghgprotocol.org/

Chapter 2
Disruption

The transition to a Green and Circular Economy and implementing sustainability will create new structures for societies and for businesses, so people can thrive in a fair and sustainable world. This fosters disruption or disruption emerges from new market potentials. New actors will find new solutions based on these new market conditions. Society must transform, and political leaders must dare to set new visions on the transition to a fair and sustainable world. This will disrupt societies, especially the old, industrialized societies. The least developed countries will develop in fair and sustainable ways. The disruption of societies especially in the countries that account for a significant part of the global economy will be met by huge changes in many aspects also from the creation of new consumer patterns and new market conditions for businesses.

Entrepreneurs will be able to overturn the big corporations. By nature, entrepreneurs have a closer understanding of the new reality at the core of their business than the companies that emerged in a completely different age, technologically and economically.

The greatest fears of managements are external, unfamiliar market conditions that remove the basis of their business models, their products, or their markets. It may be new legislation, digitalization, disruptive technologies, or rapid-changing consumer trends. The market conditions have always been changing, although in recent years the changes are appearing more rapidly. Disruption is a focus for many business leaders and is typically linked to digital and technological development. Just like in the early 1900s with the rise of industrialization, the world is again changing rapidly. Completely new business opportunities are created, and companies will look completely different in the future. They need to organize themselves differently in a new green, circular, and digital economy and because their overall purpose and raison d'être are changing. Just like 100–150 years ago and 30 years ago when digitalization started, young entrepreneurs will knock the big corporations off the stick.

The traditional business models are under pressure due to increased online communication and digitalization, but also because of rapid, accelerating technological

© The Author(s), under exclusive license to Springer Nature
Switzerland AG 2024
G. Haar, *Rethink Economics and Business Models for Sustainability*,
https://doi.org/10.1007/978-3-031-56653-0_2

development, as, for example, robotization and artificial intelligence (AI). Some say that technological development is now exponential and is difficult to keep up with. Disruption is not only driven by new technologies but also by other new market conditions that companies must deal with. Sustainability and ESG/SDGs seem to be the new blueprint for companies and are finding ground globally. New environmental frameworks set by the ESG legislation emerging in the EU and the USA. Sustainability and the transition to a Fair, Green, and Circular Economy is also a new condition that management and (non)executive boards must deal with (Haar, Chap. 6: EU regulation to a green economy, 2024a).

> **Summing up, the following main trends will radically affect the companies of the future:**
> a. Industry 4.0 with digitalization, automation, and sensor technology
> b. Biotech
> c. A Green and Circular Economy
> d. Future consumers with their new demands and patterns of consumption
> e. Purpose over profit
> f. Global shift of political and economic power

(a) Industry 4.0 with Digitalization, Automation, and Sensor Technology

Industry 4.0 embed the technological development that is of great importance to companies now. It changes production, market channels, communication, and the actual products. Industry 4.0 is also part of the solution to some of the planetary challenges. The tendency is that digitalization and a Green and Circular Economy accelerate each other. This means that the companies that have not addressed either of these two disrupters are seriously challenged in their future existence.

Humans are highly adaptable and solve many of the challenges that we face. Companies need to embed this agility.

> **Industry 4.0 covers many different elements of the new technologies currently on the market, but especially:**
> - Digitization via the Internet access and transmission of data in entirely new ways.
> - Robotization, automation, and new 3D printing technologies.
> - Sensor technologies and scanning technologies that can be used by all industries to tag goods, identify ingredients and substances in humans, and create new diagnostic tools that support product transparency and traceability.
> - AI (artificial intelligence) and machine learning will replace jobs and create new ones.

Knowledge sharing and communication have become global due to the Internet and digitalization. Now people can communicate closely with people who live completely different lives, and consumers are able to identify and understand the consequences of the social impacts from companies and governments around the globe to a greater extent than before.

The interaction of these new technologies facilitates the transition to a Green and Circular Economy. One example is the continuous development of SmartGrid to manage our energy consumption and energy supply in the future. SmartGrid links renewable energy technologies, electric vehicles, new energy storage systems, new infrastructures to create a more efficient and up-to-date supply of power, electrified public transport, and heating based on demand and consumer data recorded closer to the citizen.

Food production is also meeting the new technologies with automation, combined with sensor technologies and digital recognition. Here harvesting, watering, and fertilization are managed online. This can happen at large scale, in small-scale local gardening, and in more, or less, closed systems. This will lead to an extension of the large-scale food production, already in place on large, efficient farms of today. It also means more local, peri-urban, small-scale production of food using these new technologies. New technology available in combination with local citizens' involvement can upscale the production from local urban gardens to produce enough food to supply to urban areas. Here people can contribute and gather to form a new type of urban gardening.

This polarization of manufacturing size and intensity will also develop in industries with more local, small-scale, industrial production (3D printing). It is already happening with the global experimenting of Fab Labs. Within a few years, it will become profitable to implement productions at small scale due to Industry 4.0. This will make production more flexible and more sustainable. For the past many years, scaling has been about standardization to produce large volumes of similar products at very cheap prices. In the future, technological scaling will cause flexibility, automation, and digitalization, as well as small-scale local production, on demand. A movement towards a sustainable, order-based production of quality products, rather than a production of large series with overproduction creating overconsumption and waste of resources. The global value chains of serial production is difficult to convert and adjust, nevertheless technology is offering new and more responsible solutions. In that way technology is promoting less stock production, less overproduction, and less waste through local and orderbased production.

Unidirectional large-scale production was part of the outsourcing to Asia and kicked off the *take-make-use-waste* business model and linear consumption patterns. This must change because overproduction and overconsumption have such large environmental, social, and climate impacts. The future will require flexibility and proximity in production, and these innovative technologies will support this development.

Industry 4.0 and the new innovative technologies is a huge topic which largely affects all people and all businesses. In this book, the subject will not be dealt with in details but will be touched upon when it interacts with the other subjects. However, it is important to understand that for many reasons, Industry 4.0 is one of the great paths to a more sustainable world.

(b) Biotechnology and Genetic Therapy

In the span of technology and biology, new biotechnological methods and innovations appear. Over the next decades, we will see quantum leaps within this area. This goes for innovations within cell biology, enzyme kinetics, and genetics in completely new ways. It will affect the traditional industry and the entire healthcare field, both treatment and diagnostics at a scale difficult to imagine today. This will also affect how we clean up pollution, how we manufacture food and products, how we store information, and how we detect ingredients in all sorts of things. A lot of research is done in this field, and it will change our surroundings and the businesses drastically, especially in the new Green and Circular Economy. It is important that we dare to take advantage of the opportunities that these technologies provide. And we must dare to regulate and set industry rules for the usages of these new technologies. In this way a better and more sustainable world can evolve, and the 10–11 billion people that will soon inhabit the planet can live in sustainable and biodiverse ecosystems.

(c) The New Green and Circular Economy

The new Green and Circular Economy is a shift to a resource-driven economy supplied by renewable energy sources. It will force longer product life and durability, maintenance and repair for reuse and recycling, new clean material loops, and a completely new understanding of resource efficiency. On average, 50% of the cost base of a manufacturing company is material costs, while only approx. 30% is salaries. This leaves a great potential for rethinking resource efficiency and optimizing company costs through new business models based on return and take-back systems. In the future, we do not need more workload or a more efficient working life. We need to rethink our material resources. The transition to a Green and Circular Economy is largely driven by legislation, and especially the European Commission's New Green Deal and the Action Plan for Circular Economy set a new framework for companies and products. At the same time, legislation on products and the labeling of product environmental footprint (SPR) with Digital Product Passports (DPP), together with the new disclosure regulation on corporations, set completely new requirements for documenting sustainability, all of which has significant impact on companies (Haar, The Great Transition to a Green and Circular Economy, 2024b).

Visiting NASA in Johnson Space Center in Houston is very inspirational in understanding the need for a Clean and Circular Economy, based on renewable and recyclable resources. The NASA missions to Mars and to the Moon (Lunar Artemis Project) is conditionalized by the ability to build a habitat in an environment of scarce resources and to continuously reuse and recycle these. In this way, space missions again can set new standards for how we live and manage planet Earth.

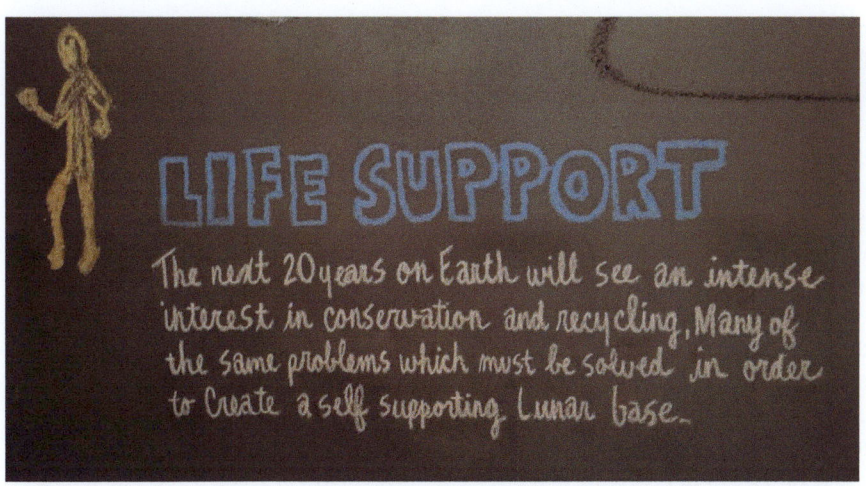

Picture from a wall in the exhibition of Johnson Space Center, NASA, Houston. https://spacecenter.org/exhibits-and-experiences/. Photo: Gitte Haar

"The possibilities are limited only by our imagination and determination, and not by the physics." Mike Duke, American Businessman

(d) Future Consumers and New Consumer Patterns

Some of the young, trendsetting consumers are very aware of the impact they cause when shopping as well as the impacts of the companies, to a much larger extent than older generations. Companies are beginning to understand the exposure they subject to by consumers. Today, companies are constantly examined in how they behave. Large global corporations have gone out of business because of single attacks on their reputation and behavior—whether justified or not. Consumers are also employees and use social media, just as new communication channels are constantly emerging.

Segmentation and targeting of consumer behavior are becoming far more individual than traditionally thought of. In the past, marketers had tools to categorize and segment consumers according to selected patterns of consumption and then target the marketing of products accordingly. It is a thing of the past that storytelling is supposed to be such a big part of marketing. Companies need to hit customers in new ways. AI and the access to consumer data are to some extent able to identify and manage marketing in new ways, but the interactions between customers and providers are changing.

The consumers and employees of the future, generation Z (Zoomers), have a different starting point than the post-war generations. They live with different concerns for the future with climate change and a poor state of the planet, overpopulation, and the struggle for natural resources. Previous generations did not have these worries. In addition, increased refugee flows and a fundamental shift in economic and political power away from the Western world and towards the East are also a new future for the very young. The post-war generations in the Western world grew up in security with increasing prosperity, democracy, human rights, and almost unlimited access to consumption and resources. Global prosperity and its distribution have increased dramatically since World War 2, yet the very young in the old, industrialized countries face a more uncertain future. This brings different consumption patterns and different expectations of companies and society. Especially the very young, as the Greta Thunberg generation, will demand unparalleled sustainability and responsibility.

The future consumers make much more use of the online market channels and the new communication platforms (ICT). They are far more influenced by SoMe than by traditional advertising. However, regulation and training in the use of online communication and data mining are important, and a more critical approach is needed to how businesses and others use data and online communication. These cause far-reaching problems that have only just emerged. The EU has embarked on comprehensive regulation of the digital companies, online communications, and media. Many of the new digital consumers know that they are making important knowledge available to the digital providers, and they want something in return. If it is not obvious which product is offered, then you and your data are the product. This needs to be dealt with in a regulatory and global perspective.

The young, impact-conscious consumers are also far more adept at using information available online to map the product value chains. They have completely different demands for transparency and sustainability of both the companies and the products. Product impact—sustainable and social—is as important as the product itself. There is a kind of common ground between some middle-aged (55+) who grew up before outsourcing and overconsumption and these young, impact-conscious consumers. These generations are aware of the necessary changes and don't just support *take-make-use-waste*. The generations in between, those now 25–55 years old, grew up in a time of consumption and unlimited access to resources and live with the perception that consumption is good and more consumption even better. These are rough generalizations and an attempt to look at different trends in society searching for groups that will bring about the change.

Information flows between communities and across borders, and consumers have online access to exactly the product they are looking for. They choose from far more *shelves* than before. Young people are subculture *shoppers* and are less concerned about which subcultures they belong to. They move between subcultures and thus also between products. One day she is vegan, and the next day she drives a Formula 1 car, EV of course. They are strongly influenced by music cultures, as seen in the 1950s–1960s, but also by cultural movements as anti-racism, MeToo, LGBT+, etc., where norms, authorities, and societal boundaries are broken down. It makes it difficult for the "old" PR folks and marketing bureaus these years to find their heads and tails in all these new trends that are not easy to segment. There is still a lot of purchasing power with old and traditional consumers. But the wave of young, impact-conscious consumers is constantly overpowering.

Storytelling and lifestyle influencers have become an essential part of corporate marketing to reach customers in new ways. This has led to a wave of pink- and greenwashing because advertisers are aware of the new demands and the new trends, but companies have not fully understood how to meet these demands in depth. Marketing legislation has been too weak to catch this pink- and greenwashing. Now there is new legislation in the EU on green claims to counter this. Companies must change their products, their marketing, and their business models to stay relevant.

(e) Purpose Over Profit
In large parts of the Western world, a dawning trend has started to showdown the *take-make-use-waste* consumer model and the old linear business models. To meet the needs of 9–11 billion people, we need to look at the global value chains and how we are using our material resources sustainably.

> The future calls for better—not more!

In the past few years, companies have been focusing on finding their purpose—and the shared values with society. Companies need to decide on the promise they make besides making money for their shareholders. The long-term perspectives of the companies are rising rather than short-term financial returns. For many years, business leaders have had a one-sided focus on customer needs, and not looking back in the supply chain, or look around in the society they share. That time is over. The long-term views and the larger perspective are important in business management of today. This is also seen in the ongoing update of the guidelines on corporate governance in the EU and globally, with demands for increased corporate responsibility in the full value chain. A one-sided focus on the quick financial gains is changing among several of the leading executives. This is also seen on the agenda of recent years at the World Economic Forum, also including global business executives. Many analyses show that companies with strategically embedded sustainably, diversity, and strong corporate governance also show higher financial returns in the long run (more on company purpose and sustainable leadership in Chap. 4).

The company WHY has been part of many strategic discussions among leaders since Simon Sinek introduced the Golden Circle some 15 years ago (Sinek, 2011). It was the big thing in the 2010s to identify company WHY based on this model. To some extent, this has led to *purpose-washing* for some leaders and companies as this very rarely changed the way the company operated. It was mainly used to engage employees and customers. The need for a change in company purpose now is based on the need for change the way we live; the way we do business; and the way we consume to regenerate the planet humans depend on. This requires new insights into the complexity of genuine sustainability, the full value chain, and the impacts of corporations and products.

(f) Global Shift of Political and Economic Power

Global power is shifting from West to the East, not only the economic power but also the political power. Wars and especially the war in Ukraine with indirect involvement of not only EU but also the USA and China have made it clear that global power is changing. The military machine of China is estimated to level the American. New parameters that are affecting businesses are today and that were not significant in corporate governance just a few years ago:

- A shift of political and military power from West to East with China
- The war in Ukraine with involvement of EU, the USA, and China
- The Taiwan conflict between China and the USA
- The opening of the North Passage due to melting of ice
- Melting of ice cap revealing access to critical minerals and raw materials
- The global financial, technological, and manufacturing dependency on China
- Global resource scarcity
- Vulnerable global logistics due to Taiwan Strait, Panama, and Suez and long, untransparent supply chains

All these parameters interact and are interdependent causing political and economic instability for societies and for corporations. They affect the political frameworks that companies are dependent on, and companies are affected both directly and indirectly. Executives and corporations are increasingly expected to take a stand. The EU and the USA are sanctioning Russia and that have had enormous consequences on companies operating in and with Russia. As an example, the Carlsberg breweries in Russia were nationalized by Putin, recently, and handed over to a Russian corporation. Before this, many European companies had to withdraw their activities in Russia and close their trade with Russian companies. Corporations have since the cold war lived good and quietly in trading and establishing businesses almost anywhere in the world. These days are over. Now the changing geopolitics situations must be considered in the board rooms instantly, and future investments must be subject to analyses of the future changes in geopolitics (more on this in Chap. 3). This introduces a new letter to ESG(G). Geopolitical for leaders to unerstand and implement.

References

Haar, G. (2024a). Chapter 6: EU regulation to a green economy. In G. Haar (Ed.), *The Great transition to a green and circular economy*. Springer.

Haar, G. (2024b). *The Great transition to a green and circular economy*. Springer.

Sinek, S. (2011). *Start with why: How great leaders inspire everyone to take action*. Portfolio.

Chapter 3
Rethink Economics and Business Models

This chapter discusses the need to rethink economics on a society level, the effects on market conditions and on consumer economy to reflect sustainability, and to facilitate the great transition to a Green and Circular Economy. This chapter also discusses the embedding into companies and their challenges in the transition and on how business cases and business models need to be understood in new ways.

About 1.5 billion people live with consumption patterns of overconsumption, wealth, and capital power. The people who directly or indirectly cause the problems have no interest in solving them in the short run. Although they make up less than 20% of the Earth's population, they are left with all the political and economic power. In the long term, it must be in everyone's interests that we all live in a fair and sustainable way and that we have natural ecosystems and a climate that allows the planet to carry the 9–11 billion people that will inhabit this planet within the next decades. Businesses play a crucial role here, and there is a lack of the knowledge necessary to understand the overall interrelations of the way a company acts and the state of the planet, to change the economic models and the business models for the benefit of all.

> It is unacceptable that the business financials are completely decoupled from the ecological carrying capacity and only has profit and growth as a goal.

New Society Model

The capitalist model and free market economy is being challenged from many sides and debated within social and economic sciences. Capital is accumulating on few hands, the increasing inequality in the Western world, and the drag on nature and the climate is now threatening human existence and our habitats.

The wealth created during human development and human spread comes from harvesting and exploiting nature. Creating agricultural lands, forestry, mining, oil extraction, urbanization, and so forth. All of these on the cost of wild nature, biodiversity, ecosystems, and the climate. For thousands of years, uncontrolled exploitation of nature has been possible because the spread of humans and their activity could be buffered by abundant nature. Nature is extremely resilient and delivers the buffer capacity to counter changes in climate and natural disasters as well as the bloom of species—plant, animals, or microorganisms. Now humans have destroyed the wild nature due to their spread and overconsumption that the buffer capacity and resilience has disappeared. This is mainly a treat to human living rather than a treat to nature itself, as nature will emerge rapidly when left to flourish.

In a world mainly driven by pure economic interests, it has become a real problem that humans can almost unaccountable harvest from the nature that we are so dependent upon. In a world of scarcity of nature and raw materials, it becomes an act of theft to destroy wild nature. It is harming other people and future generations. We need to account for the resources most dependent upon. This is not socialism. This is supporting the liberal market with an economy based on free and equal access to nature and natural resources. The free market of today does not encounter the costs or impacts on nature. Put differently, companies and consumers do not pay for the drag on externalities, as natural ecosystems, raw materials, climate, or environmental pollution from production, agriculture, waste deposits, and others. For historical reasons and because of the time slag from harvest and mining to the negative impacts on the environment set in. The same situation goes for the *new raw material*, human data, that many of the largest international corporates base their business models on. These data are freely accessible and with no control on use and spread of personal and private data. Legislation is introduced on handling and sharing of personal data, but it is still an almost uncontrolled market.

> Exploitation of externalities is often free of charge, and the pressure on natural ecosystems, biodiversity, and climate are now threating human existence and habitats.

Another problem with the free and uncontrolled access to natural resources (and data resources) is that it creates an accumulation of capital on very few hands. The first who achieve unfettered access to the resources become extremely powerful, and with no accountability or payment for the consequence on externalities. A situation close to monopoly has occurred during history and especially during industrialization. Access to raw materials has historically been created through bribing or "convincing" the way with local landowners or politicians not securing a fair and sustainable solution over time, and not ensure equal access to providers to the market of the raw material. It has just been a fight often won by the largest and most dishonest. Now it is time to ensure that invested capital and access to resources has the most sustainable impacts to most.

Accounting for the long-term impacts on the planet and humanity is lacking in almost any economic models. When economic goals are used unilaterally over a longer period, humans end up negatively impacting and exploiting our living conditions and the raw materials we depend upon. This is now happing to an extent that is comprehensive and irreversible.

The enormous global growth, especially in the Western world since World War 2, and the rapid increasing population has had far-reaching consequences for our future access to resources, land, nature, biodiversity, and climate resilience. All these elements are the pillars of our existence and have not been duly considered in the global spread of welfare. Scientists have developed a model to assess this, called the planetary boundaries—see the Stockholm Resilience Centre (https://www.stockholmresilience.org/research/planetary-boundaries/the-nine-planetary-boundaries.html) for more information on this scientific concept and on environmental impacts from human activity (Haar, The Great Transition to a Green and Circular Economy, 2024b). See the rapid exceeding of the planetary boundaries in Fig. 3.1.

In Denmark, it has been decided to develop a model that includes the impacts on climate, nature, and environment in the state economic model of the Ministry of Finance (Green REFORM) (Danish Ministry of Finance, 2023). A group of economy professors in Denmark calculated the yearly negative effects of air pollution, biodiversity loss, and GHG emissions to an amount of €32 billion for the year 2022, compared to the GDP of €375 (Sørensen, 2023). Less than 10% of negative economic impact is an odd and very low number and only reflects the negative impacts of 1 year of financial activity. A more reasonable number would be to calculate the accumulated liability from the poor state of nature, biodiversity, and climate of society due to decades of mismanagement and free exploitation of nature and the future loss of GDP from climate change, poor nature, and lack of biodiversity, like a balance sheet of a private enterprise.

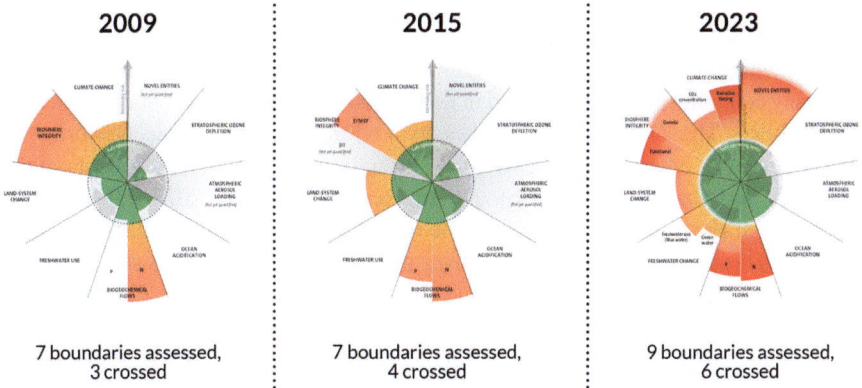

Fig. 3.1 Planetary boundaries. Illustration of planetary boundaries—a method to assess the human impact on the planetary boundaries by Stockholm Resilience Centre. Here illustrating that six out of nine planetary boundaries are exceeded

We must understand that the economic growth of the last century has come with a large price now causing lack of economic growth and large areas unfit for human living, unless nature and climate is regenerated. The good news here is that economists have started to assess the financial impacts of negative environmental impacts, not necessary the way they have approached it.

Solutions Inspired by the *Doughnut Economics*

Now facing many natural disasters created by overpopulation, human activity, and overexploitation of our natural ecosystems, we must either become drastically fewer people on the planet, or we must live completely different.

In a humanistic mindset, cutting the population drastically is not an option unless we are "helped" by unforeseen circumstances, as pandemics. Some may argue that the virus-pandemics in the last decades as MERS, SARS (corona), and Ebola are nature's way of regulating the size of human population. A well-known phenomenon in biological evolution. Recently, scientists are publishing and debating global strategies for minimizing population by lowering fertility rates (https://www.scientificamerican.com/article/population-decline-will-change-the-world-for-the-better/). Population growth has always been seen a driver for economic growth, and some scientists now argue that with the poor state of the planet now, opposite is necessary and that we need to lower population to 3–4 billion people to retain an economic sustainable future.

No matter if we half population size or not, we must find ways to live within the planetary boundaries. We also need to rethink economics and live very differently if we are to solve the challenges from climate change, natural disasters, loss of biodiversity, and resource scarcity.

Just as we must re-design products and the business models for a clean and Circular Economy, so must our overall socioeconomic thinking be redesigned into a new kind of Re-Neo-Keynesianism. This to limit the economic development that impacts our climate, environment, and draws on natural resources that are running scarce.

Many economists have speculated in new economic models that contain the cost of externalities. The *Doughnut Economics* by Oxford economist Kate Raworth has reached a lot of attention. This concept is strongly inspired by nine elements in the planetary boundaries (Azote for Stockholm Resilience Centre, 2023 (2009, 2015)) described in Haar, Part I, "Introducing the need for a Green and Circular Economy" (2024a). This gives interesting perspectives on what economic models should embed to create the change needed for a fair and sustainable planet. Figure 3.2 illustrates the doughnut model, and it operates with a social foundation (inner boundary) and an ecological ceiling (outer boundary) of the doughnut (Raworth, 2018). Within these illustrated as the shape of a doughnut, there is a range where human can exist without harming the planet or the social needs. These boundaries must not be shortfall or overshoot. The model is good and self-explanatory. It seems as if the Nordic countries have managed to create a sound and resilient social and economic foundation. Now

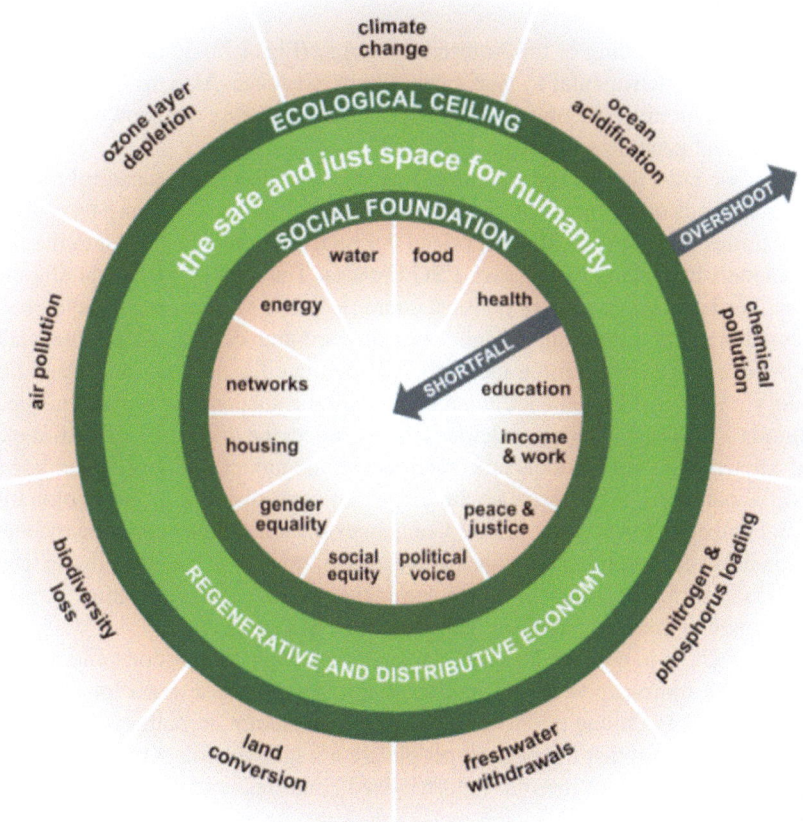

Fig. 3.2 Doughnut Economics. The doughnut model developed by Prof. Kate Raworth is based on the concept of planetary boundaries

we need to implement the ecological boundaries as the Nordic consumption, waste generation, and climate impacts are some of the highest globally.

The implementation of new economic models embedding sustainability must stand its trial. Many cities including Amsterdam (Holland) and Aarhus (Denmark) and other municipalities have announced that they will implement the doughnut model. Follow the work of Professor Kate Raworth at https://doughnuteconomics.org/.

The capitalist economy is also broken because it creates a strong concentration of capital and has done so extensively over the last decades, or maybe even centuries. History has shown that the free-market forces have led to monopolies and oligopolies. Not vastly different from the communist experiments, we saw throughout the twentieth century. A high degree of regulation and control is needed to ensure free competition, which creates a sustainable use of resources. This regulation must be guided by the principles of a clean and Circular Economy free of waste and fossil burning.

In the understanding of the Anthropocene Age (Fig. 1.2), human impacts on the planet are now detrimental to man's own future existence. Also, to the natural ecosystems and biodiversity on which humans depend to secure our food base, medical research, and general well-being.

Now, we must dare to regulate and put a price on the access to natural resources and other kinds of resources that today are harvested for free. The long-term drag on externalities that are currently not priced nor regulated must be built into the financial models by pricing of negative ESG impacts. This requires political alignment across borders, and it requires political courage. It will mean an increase in product prices, and in the short term, it might cause an economic decline. Many free marketers believe that markets force themselves to regulate for the drag on externalities. History proofs that humans are unable to create the sustainable demand needed to maintain a sustainable living and prices do not reflect the drag on externalities either in the macroeconomic models or in microeconomic models that businesses use and report accordingly. Also, because the long-term consequences of our exploitations are not embedded in product prices. Companies must have the economic incentives to transform to sustainable business, and this is now possible with the extensive ESG disclosure regulation that since can make products and companies subject to environmental taxation based on their negative ESG impacts.

By switching to a Green and Circular Economy, many of our consumer products will become more expensive. In return, products will have a much longer life and can be reused again and again and change our consumption patterns. This will to some extent ensure that the poorest people are not hit the hardest by this transition. Especially, everyone in the old, industrialized countries is to change consumption patterns and lifestyles. This does not necessarily mean that the most challenged groups will suffer because the transition means more local production, more repair and maintenance, and a large economy of used and recycled products that will support not only sustainability but also a new economy and new business models dependent on labor and handcraft skills. Although consumption of new products will be more expensive for the high-end consumers to pay for the transition, the economy of reused products will support an economy that will benefit all levels of society. Then the transition will also benefit the health and security of those who have the least.

It is important to preserve the healthy elements of the free market economy, namely, the competition between private enterprises, which provide citizens with the best solutions in the most efficient way.

> If we valuate climate change, biodiversity loss, and resource scarcity seriously, we must create cross-border pricing and regulation of access to externalities.

Creating the New Market Conditions

Now, several CEOs of global corporations are talking about a new business humanism, which puts our economic models up for rethinking. This book and the thinking here are based on well-known neo-Keynesian thinking with a liberal market economy. It is certainly not an attempt to create a socialist economy. The original Keynesian was revised to neo-Keynesian models for meeting the new global markets. Now tight regulations are needed to enter the Green and Circular Economy for capital, environment, biodiversity, and climate to create free competition among private enterprises. The cost of externalities must be paid directly or indirectly when consuming and polluting. This will lead to new business models and products adapted to the new economy.

The first step to be able to regulate and control the access to externalities is measuring and accounting the ESG impacts in the full value chain as seen with the new disclosure regulation implemented in the EU and USA these years. The strong corporate sustainability disclosure regulation (CSRD) on ESG is the starting point for accounting for the impacts. Hereby data are in place to start the pricing of and asking companies to pay for their negative impacts on the planet and humans, directly or indirectly through lack of access to the market. This will force a transition to sustainability and regeneration that humankind is in the need for. Read more about the new data in Chap. 5.

The evolution of climate change and the documentation hereof throughout the last 4–5 decades, driven by science becoming available has resulted in minimal reaction towards the needed change in the businesses. This is the proof that the capital markets are not strong and holistic enough to react on severe impacts caused by companies and from the long and global value chains. In the USA, a market for carbon removal is emerging rapidly because companies start prioritizing climate neutral as a business condition but is it not at all at the phase that is necessary to counter climate change. A lot of capital is invested in carbon removal, and from a scientific perspective, the technological state of these technologies is not at all sufficient to have any significant impact on lowering the level of GHG in the atmosphere. More importantly is the rapid transformation from a fossil economy to an economy based on renewable energy. This is the only small window left to meet the Paris Agreement, even if the access to metals and rare soil minerals is limiting the speed of installation of renewable energy. Installation of renewable technologies is driven by subsidies in both the EU and in the USA. In China it is fully state driven. So, the free and liberal market is not proven responsive to drive the necessary transition to create a sustainable planet without strong help from regulators creating the new market conditions, and this again must be linked to new economic models to introduce long-term consequences and drag on externalities into to the markets—both the capital markets and the consumer markets.

Right to consumption is a misconception and accelarating consumption as the driver for wealth even so. Nevertheless, it is articulated as a right and a necessity by politicians and of course the consumer brands making consumption beatifying.

Companies exploit the access to consumers in new ways by social media, and they exploit the almost unlimited access to consumer data available from our online activities. This is done through influencers or by targeting the information we encounter online via search engines: AI as provided by Google, Meta (Facebook/Instagram), and others. It is not a truth that everyone has the right to consumption from all shelves. The creation of overconsumption is now at a level that the planetary resources cannot meet. Especially the consumption of products or services with a very short lifespan and a large negative impact on people and the environment from *take-make-use-waste*. One or two decades ago, only very few people flew regularly, and it was common sense to take care of your things and make them last. Today, it is the buying experience itself that is cultivated to satisfy the consumer, and thus it has become a kind of "right" or a widespread hobby in the Western consumer economies. Again, it's a misconception that consumption is a human right. It is not a human right to be able to buy a lot of stuff manufactured irresponsibly and waste it again after a few times of use, as many do with their clothes in the Western world.

> Consumption is the engine of the global economies, regardless of the impacts on planet and people in the global value (supply) chains.

The hunt for cheap prices has thus become a major challenge, and we again need to understand that quality has a higher price and that the drag on planet and people from the manufacture of products must also be paid. If things become more expensive, we take better care of them and have expectations that they have a long lifespan and can be repaired and resold—instead of just being thrown away.

Prices of building materials, furniture, electronics, and many consumer goods have dropped dramatically over the past two to three decades in line with technological development and outsourcing to and competition from production in Asia. This is also shown by the lack of economic development of the general population that has been in these countries. One might expect that outsourcing would have created rising wages and wealth locally, and it has to some extent. But the spread of wealth has not happened in many places—on the contrary, the prices of consumer goods in the Western world have just fallen. Wages in China have risen, and China's economy has benefited from being the world's center of production, although China still holds a very high Gini coefficient and high inequality. Now the productions are being moved to other low-wage areas, like Bangladesh, Vietnam, etc. China has set targets to overcome its pollution challenges and wants to get rid of its dependence on fossil fuels. They now have a target on climate neutrality in 2060. But low-wage production has moved elsewhere, and the Chinese today live to a greater extent on exports of their own products developed with knowledge built up and learned from blueprints of Western technology products. LED, solar cells, and electric cars are examples of this.

The consumption economy is an economy only in place in parts of the world. Most of the world's population does not have this enormous consumption and overconsumption of goods and cause far less impacts on the climate and the environment. Approximately 60% of the global population live under the recommended climate impact per capita due to their low consumption of goods, food, and energy. The resources, wealth, and benefits available in a global perspective must be distributed globally among more people to meet a sustainable development. This inequality will push the corporate responsibility by future consumers and legislators. It will make the need for new consumer patterns important in the Western world, and this transition must be driven by the corporate responsibility of businesses.

A communication has emerged among citizens and politicians that when the Western world is to limit consumption, the poorest part of the populations here will suffer. This is a strange and short-sighted angle of a problem that challenges for the weakest and low-income in society. Those with few human resources and the low-income will largely face the consequences of climate change, lack of access to material resources, migration of climate refugees, and the lack of competitiveness that European companies have today. These threats need to be addressed in a way not leaving the weakest behind. Therefore, society models and business models must be rethought; otherwise the long-term consequences are enormous.

"Everyone wants development—nobody wants change"—Søren Kierkegaard, Danish philosopher, 1813–1855

Business Perspective

The overall need for change is obvious if the planet is to provide for all, and especially if the Western world is to maintain prosperity here. The transition to new economics and new market conditions is a new imperative for businesses, and it will be the largest changes that companies will have to encounter since the beginning of the industrialization. Companies and the business environment are the wheels of driving the chance and to be part of rethinking the economies. As a business leader, it is important to understand the complexity that not only sustainability brings, but also the shift of geopolitical power and the resource scarcity will result in a completely new economy that business leaders must adapt to.

In a new economy, companies will need:

- To be part of a new economy and new market conditions where the utilization of external resources and other externalities as climate, clean water, nature, and decent human resources will be priced in new ways. This will put the responsibility of the impacts caused on the entities that causes climate change, pollution, demolition of nature, and ecosystems and make irresponsible use of human resources.

- A new purpose focusing on the impact of their product and services in the full value chain meeting the Sustainable Development Goals (SDGs) as the world's most important plan.
- Develop a new strategy that build on sustainability and a new business model that compile with circular principles (strategy process on transforming).
- Organize themselves differently to meet legislation in a resource-driven economy and the transparency and traceability of products and materials that EU will require. A switch in primary and secondary activities of the company (Porter).
- New non-financial business data will be required from companies at the same level as financial business data—internally and externally.
- New strategic tools and ways of assessing business risks and business cases for prioritizing and selecting the scope of the company's space in a Green Economy.

This book addresses these issues and provides tools, methods, and experiences to assist leaders to drive the transition.

New Business Models and New Business Cases

In a new economy with new market conditions where companies are to transform to a Green and Circular Economy, it becomes important for management to base their decisions on new assumptions and to understand the business models and business cases in new ways. The rethinking of economies and the new society models must be brought into the companies and the way they navigate, manage, and communicate.

This is necessary when deciding on new business models within the companies, when entering new strategic partnerships to manage the new circular value chains and to attract new investments. Capital must be relocated, long-term investments must be made, and external capital must be provided for investments in new circular business models, energy optimization, and renewable energy and managing the sustainable value (supply) chains, upstream and downstream to regenerate ecosystems. Reporting and monitoring of ESG data to manage the companies and to attract capital will become as important as the financial data.

> **Raising and allocating capital for the transformation of the company often comes with several challenges:**
> - Financial partners as investors and bankers are often not familiar with the business models in a Green and Circular Economy. When rating a company, investors look at historical data and traditional market analysis based on traditional business models, rather than the future market conditions and the new circular business models.

- Investments in renewable energy, energy optimization, and energy conversion (electrification) often hold a high initial investment and lower operating costs, because the energy source is free and inexhaustible (wind, solar, water, etc.). Often only the initial price of the installments is subject to evaluation and tenders, and a simple payback time on the initial investment is calculated. Many companies have investment criteria of 3–5 years payback time. The long-term, reliable investments that create resilience, as renewable energy, are not sufficiently valued.
- There is still political instability around the pricing of access to renewable energy and access to the grids, due to inconsistent and rapid changing public financing and promotion schemes. Fossil fuels are still strongly supported by public funds, and the pricing of access to RE are considered more uncertain than the public financing and supported fossil fuels.
- Management and financial partners are not used to monitor and estimate the long-term consequences of ESG impacts. These years' reporting standards are implemented in the EU and in some states in the USA, and this will provide the data needed, but competences still lack to use the data to drive impact changes required by the market and by (future) legislation.

If the business cases are to prepare companies for a Green and Circular Economy, the way investments are accessed needs to change. Financial partners and investors very often base their assessments on historical data and traditional business models from a linear economy. This is a challenge when an existing company transforms its business models into a new and more resilient, green, circular business model and long-life technology. It is often difficult to raise capital for the transition, especially if the company historically has been challenged by the old business model and has weak financial records.

The best solution here would be to educate bankers to understand the Green and Circular Economy and the great transition creating new market conditions. This is quite a task. Typically, bankers are conservative, and these years they are busy getting their own businesses on track, sustainable, responsible, and compliant. After the financial crisis, the reputation of the banking sector is poor, and their attentions are on countering digital fraud, anti-money laundering, and tightened legislation.

In the EU and in the USA, different types of support and green investment funds have been established by the governments to finance the green transition. Often capital is still mostly granted to energy renovation, renewable energy, RE technologies, and to a lesser extent the transition of business models to a new economy, new market conditions, and circular business models. Green Fund capital is supposed to be risk-averse and should to a larger extent transform existing business to new circular business models, as the renewable technologies are becoming profitable. EU Horizon funds are now investing in new circular projects and circular business models, but there is still a large potential in transforming the existing industries to a Circular Economy. Energy optimization and renewable energy is profitable for the

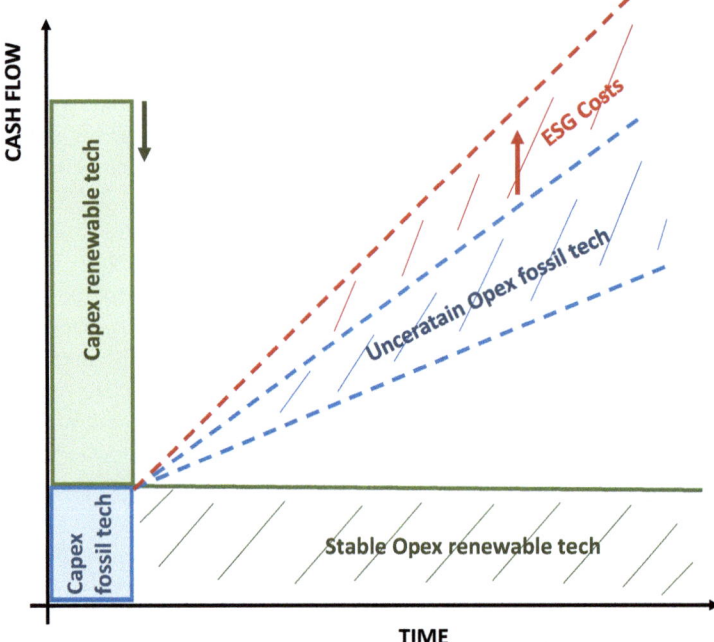

Fig. 3.3 Shift in business case. The overall difference between a business case of fossil technologies and renewable technologies. The initial investment (capex) of the renewable technologies is larger than the fossil technologies. Whereas, the operating costs (opex) of the renewable energy supply are low, and since the source is abundant and free, the cost will continue to be low and stable. The costs of fossil energy sources will be continuously increasing and uncertain in the future, also due to ESG costs from negative impacts as carbon tax, higher interest rates, and difficult market access of the old technologies

companies, whereas Circular Economy is about creating new business models and new products to meet new market conditions, and this will require capital in the existing industries if to accelerate rapidly enough. The new ESG reporting directives (SFDR and CSRD) will hopefully promote these investments and measure the investments on other parameters than financial returns and increasingly also on especially the new environmental parameters in the sustainability standards: climate change and adaptation, Circular Economy, biodiversity, regenerative ecosystems, and no-pollution.

When companies invest in renewable energy or electric cars some, banks provide lower interest rates for RE and EV. Still there is a need for implementing these technologies at a greater speed. The technologies are still new and large investments that will not pay back within a few years. The business cases (BC) for RE technologies or EV offer lower operating costs (opex) and a high initial investment (capex) compared to traditional fossil technologies, as illustrated in Fig. 3.3. Renewable energy technologies often hold a long lifespan of 20–40 years, often longer than many fossil technologies. Therefore the business cases must be understood in new ways. The energy itself is free, as access to wind, sun, and water is abundant and accessible to everyone. Harvesting it and turning it into electricity carries a cost, though it is a predictable cost for the company, contrary to the prices of fossil fuels.

Maintenance costs of renewable energy technologies are typically much lower than for combustion technologies, which is a complex technology and difficult to maintain due to wear and tear, fuel pollution, etc. Combustion technologies are widely used both in heat, electricity, and propulsion (transport). The benefits of the longer lasting and cheaper renewable energy technologies are left out when companies use the simple payback time as an approach. Many companies have a policy that new investments must be paid back over 3–5 years from the returns they earn. Then an attractive business case as the investments in renewable energy supply comes short and the installment of renewable energy will not happen rapidly enough and the upside of securing stable and cheap energy costs in the future is overlooked. Traditional BC approach does not encounter the stability of future operations, the operating life, and the positive ESG impacts as, for example, the potential for circular material uses and reuse.

The fossil technologies deliver an operating cost that will increase and be unpredictable in the future, due to varying market prices and due to increasing carbon and environmental taxes. This is not very attractive for companies that look for safe, stable, and predictable prices on their energy supply (opex). See Fig. 3.3.

The storage of electricity is still a technical challenge, but much can be done by combining various RE technologies and integrating with waterborne and/or ground heat-HVAC. Solar heat panels and geothermal installations are also renewable technologies creating heat or cooling that must be included in the system to meet energy demand 24/7. As an investment case, renewable energy technologies are more profitable and safer, long-term investments than combustion technologies. For a certain period, many countries promoted investments in RE through state-financed support schemes and guaranteed settlement prices for power from RE plants. Now the global demand for RE (solar and wind) cannot be met by the global production capacity because China and the USA are installing at high speed. Large corporations and even small private enterprises must consider RE investments and installations locally due to the very attractive return on investment (ROI) over the lifespan of the technology. Learn more about electrification and renewable energy in another book of the same author (Haar, *The Great Transition to a Green and Circular Economy*, 2024b).

Not only is it necessary to approach business cases differently; it is also necessary to understand the financial effects of the new circular business models to promote investments in these. The circular business models are often unproven, and even so it is important to be able to attract capital to the circular transition and not only investments in RE.

Figure 3.4 illustrates the need for understanding the difference in business cases in a fossil, linear economy, and a Green and Circular Economy. The traditional BC with short payback time on investments as known in a fossil and linear economy foster short investments and short time for return on investment.

The circles in the figure indicates the size and level of profitablity, and as seen the Green and Circular Economy provides a larger financial return that the fossil and

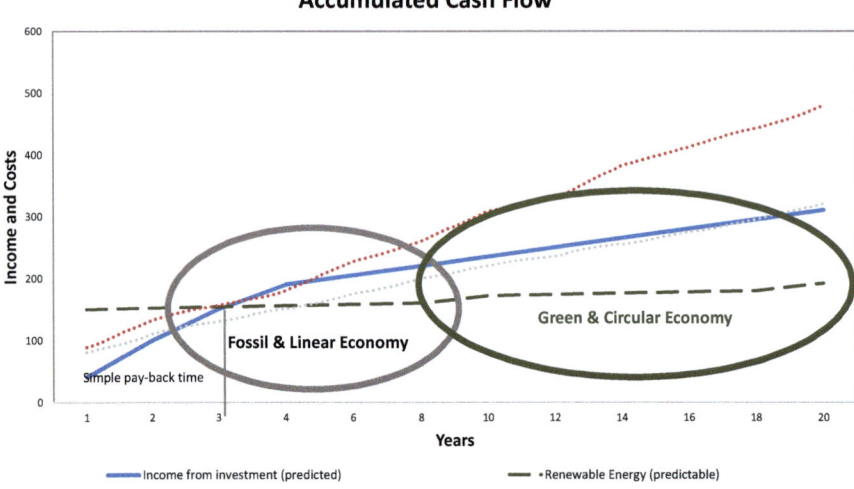

Fig. 3.4 The new business case comparing linear and circular business models. The figure illustrates the difference between the BC in a fossil and linear economy, and the BC in a Green and Circular Economy

linear economy. This is due to rapid changes in the market situation and with short-term business models based on products with short lifespan. The costs in the fossil economy are more unpredictable, due to rapid changes in raw material prices, as fossil energy, virgin raw materials, and not the least the increasing ESG costs to be charged in the future, as carbon tax, higher interest rates, and decreasing market access. The BC in a Green and Circular Economy needs time to pay back because the initial investments are higher. On the other hand, the costs of the Green and Circular Economy are predictable since it is based on inputs that are in abundance, as renewable energy and recycled materials.

The fossil and linear economy will soon be subject to market restrictions and environmental taxes to minimize the drag on the planet, as well as increasing capital costs. The regulations on ESG disclosure are the first steps towards an economy where companies must pay for their full use of externalities in the extended value chain.

> When evaluating the business cases of renewable energy and the circular business models, the unknown ESG costs must be a part of the financial evaluation of a company. They will only increase in the future.

Understanding the differences in the business cases and circular profit models illustrated in Figs. 3.3 and 3.4 must spread to management. Leaders must transform and understand the business models to meet a future Green and Circular Economy with a different cash flow that requires time, patience, and visions for the future to save the planet and to provide financial sustainability of the companies. This is how the rethinking of economics will directly affect the businesses.

Plenty of power-consuming industries needs to electrify and shift to renewable energy. It should be possible to obtain attractive financing for this in a financial market looking for long-term investments and with tightening ESG requirements. Otherwise, the levelized cost of the power from RE are now cheaper than any fossil input, and since the war in Ukraine, even coal is more expensive than RE. The renewable energy technologies as solar panels (EV), wind turbines, and batteries come with significant warranty periods from the manufacturers, and the real lifetime of these technologies are significantly longer than the guarantee period. So, when EV are projected with a lifetime of 25 years, most EV panels maintain a reasonable performance for up to 40 years, and then they must be recycled into new EV panels in a circular business model. The same applies for wind turbines. Electrification means energy savings, as the combustion technology is a very energy-inefficient technology compared to electrical power supply. There is a tremendous loss of heat from the combustion engines (ICE) and thereby a direct loss of energy of up to 50%. So, electrification of industrial processes will not only facilitate input of green affordable power supply (RE), and it will also save large amounts of energy (Haar, The Great Transition to a Green and Circular Economy, 2024b).

The Electrical Vehicles as an Example

The electric vehicle (EV) is a good example of a new technology providing a better business case than the fossil technology and the combustion engine. Still, many restrain the EV, even if it is easier, cheaper, and more fun to ride an EV. The battery technology and range still create uncertainty, and the governments are not prioritizing the transformation to electrical transport with the necessary incentives and charging infrastructure.

The electric engine is a simple and old, well-tested technology that does not require nearly the same maintenance as a combustion engine (ICE). The energy consumption is almost twice as high for gasoline as for electricity for the same range, due to the heat loss from the combustion process. The energy consumed to transport electricity (line loss) is substantially less than the loss of extraction, refining, and transport of gasoline. Electricity has an approximate utilization rate of more than 80% from the source of production. Whereas, gasoline/oil has a utilization rate of only approx. 40%. The huge difference in efficiency shows that no matter what source of electricity, the EV is a far more energy efficient technology and must be the technology of the future.

This is true especially now that the Li-battery technology is offering attractive ranges of the electric cars. It is the batteries in the electric car that is the new, expensive, and heavy technology. Therefore, manufacturers typically guarantee battery performance, and batteries may be replaced when performance falls below approx. 80%. This provides the EV with a longer lifespan than ICE. The electric vehicles are thus much cheaper to maintain, and the critical parts (batteries) are recyclable and upgradeable as the capacity of the batteries increases—happening very rapidly now. So, if the bodywork of the electric cars is designed and manufactured at good and circular quality, the engine and bodywork will last for many years, and the batteries can be replaced repeatedly.

Examples of Tesla's that have driven 1.5–2 million kilometers are out now, though with replacement of the batteries and for some also replacement of the engine. Therefore, electric cars hold a much longer lifespan than most combustion cars put on the market today and with a much greater potential for maintenance. Exactly what is required in the Circular Economy.

Thus, the depreciation on electric cars should be much lower than on combustion cars, why electric cars now are subject to more attractive financing. Now almost all car manufactures provide good EV models, and new car manufactures are appearing with Tesla as a global leader that soon may be overtaken by Chinese car manufactures. There is still a need for fleet owners to get electric cars out on the roads and gain some experiences and some data to support the new business case. The Li-battery like the electric car also has an exceptionally long life and a large reuse and recycling potential. New battery technologies are appearing and will offer even more efficiency and range.

Just to eliminate rumors of the potential environmental impacts of the EV compared to the combustion cars, Fig. 3.5 illustrates an LCA-based comparison (Bieker, 2021).

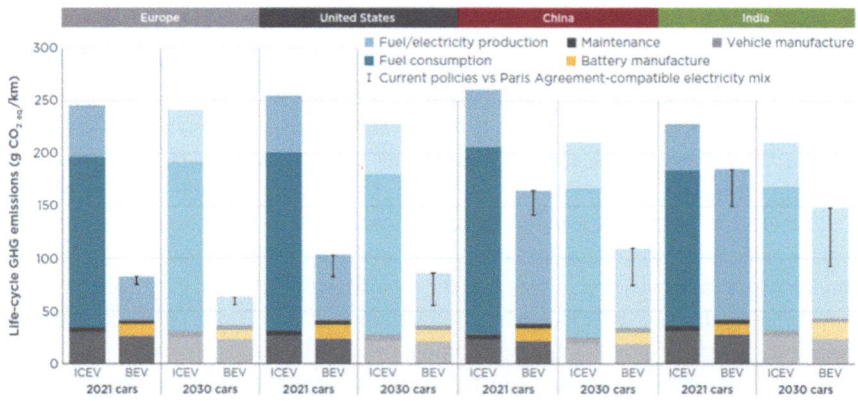

Fig. 3.5 Global comparison on GHG LCA of combustion engine and electric passenger cars. Lifecycle GHG emissions of average medium-sized gasoline internal combustion engine (ICEVs) and battery electric vehicles (BEVs) registered in Europe, the USA, China, and India in 2021 and projected to be registered in 2030. The error bars indicate the difference between the development of the electricity mix according to stated policies (the higher values) and what is required to align with the Paris Agreement

Here it is clear that it is the fuel consumption and production that emit the most GHG in both types of vehicles and that the input grid mix of electrical power has a high potential to minimize the GHG impacts from electric vehicles (BEV's) over time. The GHG emissions from the battery production are insignificant and will be even lower when the batteries are recycled. Other environmental impacts from battery production and extraction of minerals have been in the center of the discussion on BEV. It is an issue that extraction of minerals causes environmental disasters, and it does so in the production of all types of batteries. Since the batteries of smart devices are to a very low extent recovered and minerals here not recycled, it is here that that the environmental issues are now together with the environmental impacts from the lead-based batteries that are in the ICE. The high price and demand for Li-batteries in BEV is driving the shift towards more ecofriendly extraction and creation of circularity not the opposite as many believe.

Renewable energy is mostly electricity-producing technologies, and the need for battery storage—inside and outside cars, houses, etc.—will only increase as the transition proceeds. Add to this that the BMS (battery management systems) of electric cars are now able to provide electricity back into the grid, which means that cars and Li-batteries are an indirect storage for renewable energy. Read more about the transition to a Green and Circular Economy in another book by the same author (Haar, *The Great Transition to a Green and Circular Economy*, 2024b).

The electric car example is a great example of the new technologies in the Green and Circular Economy as they have longer life, better reuse, and recycling potential and are an important missing link in the transition to a fossil-free society.

The same benefits must be outlined for really many of the new green technologies that are part of the green transition and which oddly enough are not just implemented in a hurry. These new technologies are simply far more attractive investment assets and far more stable, and safer, too, both financially and operationally. One of the biggest challenges for the existing leaders is both the lack of understanding the fundamental difference in technologies and the radical change that many of these available technologies will create and what challenges they are already able to solve.

So, again, it is the fear of leaving the old shores that limits the possibility of reaching new shores. Many of the leaders of today have grown up in the fossil linear economy, and they have seen these business models work. Now they dare not face the changes that are so different, nor face the fact that they need dare to think and believe in something completely different, also when assessing the financial impacts.

Many companies find that sustainable and circular business models are more profitable than the old linear models. So, sustainability is and will increasingly become a profitable and necessary competitive edge for many companies. Sustainability is no longer opposed to making a profit. There is a rapid regulation of products and companies due to many years of negative impact on the environment. This is also facilitating the development of new sustainable solutions. Therefore, sustainability and business are now no longer opposites. They go hand in hand.

"We cannot solve our problems with the same mindset that created them."—Albert Einstein (1879–1955)

It is inspiring to consult Einstein's small philosophies when becoming uncertain with all this new thinking. Although he lived in another time with other big changes and important technological developments, his words of wisdom may come in useful when facing the challenges of our time.

References

Azote for Stockholm Resilience Centre, b. o. (2023 (2009, 2015)). *Planetary boundaries*. Persson et al and Steffen et al, Stockholm, Sweden.
Bieker, G. (2021). *White paper*. Berlin: International counsil on clean transportation. www.theicct.org.
Danish Ministry of Finance. (2023). *Climate and green economy*. Copenhagen, Denmark. https://fm.dk/arbejdsomraader/klima-og-groen-oekonomi/.
Haar, G. (2024a). Part I: Introducing the need for a green and circular economy. In G. Haar (Ed.), *The Great transition to a green and circular economy*. Springer.
Haar, G. (2024b). *The Great transition to a green and circular economy*. Springer.
Raworth, K. (2018). *Doughnut economics*. Cornerstone.
Sørensen, P. B. (2023, January). *Green breakthorugh: New Calculator puts a number on nature and climate impacts*. Retrieved from University of Copenhagen: https://www.econ.ku.dk/nyheder/alle_nyheder/groent-gennembrud-nye-regnemaskiner-saetter-tal-paa-vores-natur%2D%2Dog-klimaaftryk/

Chapter 4
New Company Purpose with Sustainable Leadership

Market conditions are changing rapidly, due to disruption, technology, online market platforms, sustainability, and new types of consumers. These are all topics that change the landscape for businesses and challenge leadership. Companies need a new approach to the outside world and their stakeholders, if not to be disrupted or dismissed. This is topped by an increasing focus on fraud in many sectors which was an important root cause to the financial crisis in the 2000s, and sustainability has also been a frame where much fraud and dishonesty have happened from the business side.

All this drive new business opportunities and call for a greater symbiosis between civil society and businesses. Some call it impact-companies. The great transition to a Green and Circular Economy and the greater awareness on sustainable business models will drive innovation and hit all companies that want to operate in the future.

A new period of social awareness or a new formation of society has started. For a long period, companies only focused on generating profits and capital growth for its shareholders. Companies are now entering into a new role in the society. Again, there will be focus on how companies behave in society, towards their employees, customers, suppliers, and other community stakeholders. During the corona crisis, it became clear that the rapid spread of the virus and its consequences pointed at the global value chains. This visualized our dependence on the global value chains and their vulnerability. How quickly and easily we can affect each other and how dependent we are on things not produced close to us. It became clear that we lack the control necessary to handle critical situations. The corona pandemic has also changed the awareness of human vulnerability and linked this to the global value chains and to our somewhat unhealthy proximity to animals. Both in the spread of the virus and in the lack of local supply of medicine, protection, and medical equipment. The corona virus has in some strange way also pointed at the human weakness, overconsumption, and overpopulation, because of the lockdowns and the sudden proximity to the people close to us and put our personal values in a new perspective.

© The Author(s), under exclusive license to Springer Nature Switzerland AG 2024
G. Haar, *Rethink Economics and Business Models for Sustainability*,
https://doi.org/10.1007/978-3-031-56653-0_4

All these trends and challenges call for a shift in the overall company purpose and missions. After World War 2 (WW2) and during the unrest of the 1950s to 1970s, a polarization emerged between political and intellectual powers on one side and business on the other side. Businesses and civil society were construed as counterparts with conflicting interests in the sense that some were fighting for distribution of prosperity and others for securing capital, as a growth engine. This is changing, and it becomes a growing tendency for companies to have a different set of values, seeing their employees and customers as immaterial capital and taking a larger responsibility in society.

Companies face new generations of consumers and employees with new expectations for companies, products, and leaders. The young generations are much more aware of their impact on the environment through their consumption patterns; and they are good at finding information about the global value chains. There is a movement in the very young generations with a greater awareness and more insights into the challenges of the future and the problems created by a global economy. Probably

New trends contributing to the changing societies and businesses are as follows.

- Global online knowledge sharing and communication—people come close to people who live completely different. More than earlier, people understand the consequences for the individual of corporate and government actions.
- Sliding boundaries between work and leisure. More people seek fulfillment through work and not through political activity, as in the previous century. The political polarization is changing as society and prosperity change.
- Young and trendsetting consumers are becoming more and more aware of the impact their consumption has on the planet and of how companies behave in the global value chain. Companies increasingly understand their exposure towards consumers and are constantly rated on how they behave. Large corporations have lost their business overnight because of breaches in their reputation and lack of reasonable conduct—whether it was reasonable or not. Consumers are also employees, and they use the social media all the time with new communication channels emerging.
- Business models under pressure because of digitalization, new technologies, and online rapid communication. Business models are also challenged due to technological innovations. Some describe this development as exponential, i.e., new technologies lead to new developments, etc.
- The Green and Circular Economy, with circular material loops, where resources are reused and recycled. As well as the switch to renewable energy sources facilitates new business models.
- A resurrection of new types of communities because the worship of the individual has shown its downside, and some of the greatest challenges in the industrialized world are loneliness and stress.
- Working remotely is a growing tendency that breaks boundaries and creates different relations and commitments between employees and employer.

because they have access to faster and more information online, and because they communicate directly with young people all over the world with completely different living conditions. The Arab Spring movement, the women's movement in Iran and the climate activism all over the globe from the uprising of young people, was made possible because of social media and the online access to communicate and to interact.

These youth movements demand dramatic changes and are driven by forces other than early rebellions. In a different way than the rebellions of previous decades, young people of today articulate that the changes are driven by economy and that an economic paradigm shift is needed because prosperity is now facing scarcity and inequality. The movements of the past created a polarization between businesses and political idealists as with the youth rebellion of the 1960s–1970s. Young people today have a different and more concerted way of demanding change. They to a larger extend see the businesses as facilitators of the changes and ask for a greater purpose of companies for engaging as customers and employees. Not only the way the young and future consumers look at companies is changing; also the way companies attract, retain, and manage the future generations is changing.

The financials and the market opportunities of businesses are affected by the same shortcomings that the young rebels worry about, as resource scarcity, food shortage, energy scarcity, nature scarcity, and skill shortages. Future businesses and future generations share values, and the overlap is large between the change that companies need to implement and the change that young people are demanding. Something similar was at stake in the youth rebellion of the 1960s–1970s. The difference is that the youth movement of today is supported by science. Young people's desire for change is not a hippie movement but driven by a scientifically proven necessity for more sustainable planet and fair living (Haar, The Great Transition to a Green and Circular Economy, 2024b), also reflected in the Sustainable Development Goals (SDGs).

Many of today's business leaders grew up and were educated in world of abundance and the dawning globalization, young people of today grow up in uncertainty, and fear of the planet will be able to provide for them in the future. This now creates a gap between how businesses are managed today and how the future employees and consumers are motivated and consume. Business leaders must understand this change of mentality and the values of their employees and customers, to transform their businesses in time to stay relevant as a workplace and in the marketplace. Today, some business leaders still make fun of sustainability and the woke movements. This is a rather dangerous approach for future businesses. A future with shared values, spread of prosperity, and a Green and Circular Economy.

It is a sound principle in a liberal market economy not to harm others through economic activities, and a principle of polluter-pays also coherent with a liberal market economy. Today, companies indirectly harm people and the environment from their environmental, social, and governance (ESG) impacts along the global value chains and have major effects on people's lives and on nature. Even if a company operates out of a country with strict legislation on labor rights and the environment, they now also need to know their indirect impacts throughout the value chain in scope 3 (UN GHG Protocol). This is expected of them and required by the new EU legislation on extended responsibilities (Haar, Chap. 6: EU regulation to a green economy, 2024a), as well as in the US with the Inflation Regulation Act (IRA), and the new Executive Order on revitalization of nature and the access to nature for all.

Purpose Gap

Plenty of new businesses and entrepreneurs meet the consumers and the employees of the future. But the large global corporations will be challenged on their business models if they do not understand the depth of these new trends. The transformation of a business needs a top-down approach based on bottom-up knowledge and starts with the business leaders. Business leaders need to focus and understand this purpose gap. Business leaders of the future have started to talk about new shared values in companies. New personal values among business leaders of the large corporations that want to survive are expected, but there is still a gap to be closed.

The purpose gap illustrated in Fig. 4.1 outlines some of the differences in the values of the young people fueling future trends on one side and the business executives on the other side of the gap. This purpose gap may put some businesses into the grave. Companies must adapt to a new Green and Circular Economy, a new digital world of Industry 4.0, and a new set of shared values creating a new business purpose.

The purpose gap in Fig. 4.1 shows the paradox between those who manage businesses today and those who represent the future market as well as the future employees. Traditional business leaders have grown up with the business model of

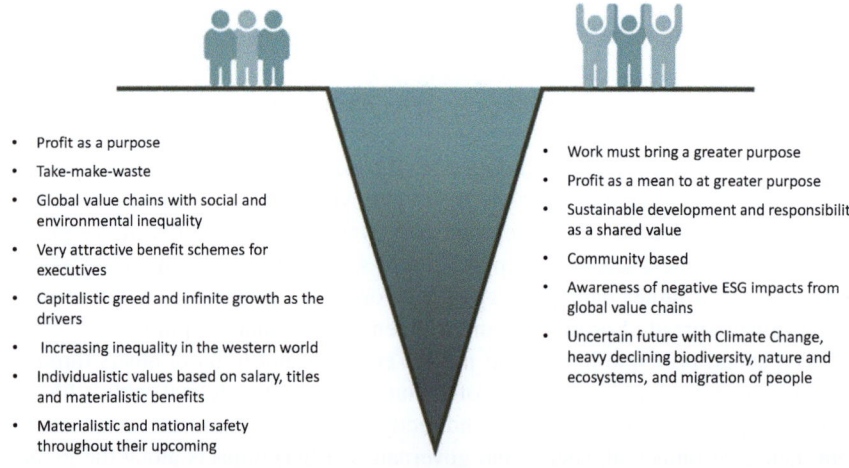

Fig. 4.1 The purpose gap. The figure illustrates the big gap that many business leaders are not aware of and that they need to understand to meet the employees and customers and bring their companies into the future

take-make-use-waste with little focus on impacts from the global value chain emerged over the last decades. Little focus has been on the actual consequences of optimizing and minimizing through outsourcing at lower costs. This business model impacts the places of sourcing and creates an unrealized vulnerability of supply to the receiving companies. Many global brands and international corporations are not even able to account for the complex value chains. They are becoming a smaller and smaller part of these value chains. This has resulted in a huge gap between the way business leaders thrive and what the future consumers and employees expect. This gap exists on a personal level as well as on a company level and on a society level. In the future, leaders (business and politicians) will fall into this purpose gap not realizing what hit them and their businesses.

In the old, industrialized countries, inequalities have increased in recent decades. Especially in the USA and in Great Britain, inequality is larger now than the last 50 years, creating political and social polarization that has become very pronounced. This inequality accelerates the need for change even further, since it has become obvious to many that the development is going in a wrong direction. When very few people become very rich within a decade and own a significant share of the global assets, the question must raise: Is this based on fair and decent growth, or is this growth created on behalf of other people, nature, and climate? It begs the debate for the need of regulation and new form of taxation to control inequality and accumulation of capital, even among business owners with a liberal economical stance and proponents of the free market dynamics.

Executives in the larger corporations are rewarded in completely unreasonable ways with benefit schemes that do not reflect the risk of such a job. A common sound economic principle is that a great financial gain must reflect a great financial risk. It is fair that extraordinary scientific or technological developments are financially rewarded when creating something groundbreaking. A CEO position in a large corporation is a huge workload and comes with huge responsibility and of course must be paid well. However, it is completely unreasonable for the CEO to become a billionaire from running a business as an employed leader. Especially when the business models are based on untransparent value chains and the exploitation of other people, nature, and climate. Today, management compensation schemes are disproportional with the average payroll in the company and in the value chain. The values created in the companies are created by an interaction between skilled employees, maybe groundbreaking technology, long-term execution, and a good go-to-market strategy. Therefore, it seems disgraceful that such large payroll disparities have occurred in companies. In the Nordics mangement compensation schemes are attractive but not at the level seen in other parts of the world, especially in the US.

The essence of the purpose gap is that profit is no longer a goal. Profit is a mean to achieve a greater purpose. Corporations must find a greater purpose and shared values with society. They will then use their market power and economical power to drive change and to leave a positive footprint on society and the planet. Companies must change from avoiding doing harm, delivering a product and jobs, to being an important part of the great transition towards shared values with society.

> Profit becomes a mean and business purpose becomes the goal.

Sustainable Leadership

There is a need for leaders—business leaders as well as political leaders to change their ways of conducting leadership as well as management, and to approach their positions differently. The Sustainable Development Goals (SDGs) and the rising legislation on ESG deliver a new framework for common understanding of the challenges and shared values of society, businesses, and citizens.

This is putting an increasing pressure on leadership in the future. Leaders must expect a shift in corporate entitlement and delivering the creation of a genuine purpose. Where making money becomes a mean to fulfill a greater purpose and regenerates a sustainable planet. A change in corporate missions. The EU legislation has extended the responsibility of company leaders to an extended responsibility in the full value chain (scope 1 + 2 + 3), and this requires new leadership skills and new personal skills.

The transition requires new corporate strategies, new management styles, and probably also new management types. Rethinking economics described earlier in this book and the transition to the Green and Circular Economy brings about a completely different way of doing business. This offers completely different goals for creating a sustainable world. We are facing a shift in the management culture with requirements of new human values in strategic management. Corporate survival is dependent on an insight into a changing world and a new economy. Management must navigate in a market where the greater purpose becomes important and keep focused on the strategic importance of the positive impact that companies must create.

The increased focus and legislative requirements on corporate purpose and on creating a positive impact on the world also mean a shift towards more feminine values in management. This is to meet the new demands from the outside world, customers, and investors and especially from employees. That does not mean that the leaders of the future cannot be men. Even today, there are many men who possess feminine values and who show much more of themselves as human beings. But it means that we will probably see greater diversity with more women in the executive management of the companies. Companies can only deliver a full-fledged strategy and execution of values when they reflect their customers and employees.

A person who contains commitment, dilemmas, emotions, humanity, and drive—leaders must show what they are made of. Such leaders exist today, and they are often successful leaders and entrepreneurs, courageous and passionate people. There is still a tendency for leaders of large corporations to be as smooth, correct, and anonymous as possible in their relation to the outside world. Because their personality or personal opinions must not stand in the way of the company brand. These leaders also tend to be portrayed—or have portrayed themselves—as very infallible and anonymous in their political views, although they are often very charismatic personalities who are good at engaging people. Business leaders have for long not been allowed to hold political opinions or otherwise make the company vulnerable. They were meant to be a statue of the image of the company, regardless of what was going on inside the company. That era is over. The young employees and consumers expect to see real people if they are to dedicate themselves and their time to the company. They want to see the CEO's family, passions, dilemmas, and desire for a better world.

"The best things in life aren't things."—Street Art

Knowledge Gap

Not only the purpose gap challenge companies and society, also a huge knowledge gap is a barrier for the transition. Sustainability requires a new understanding in all aspects of operating a company. It is not solved by employing a (female) sustainability manager or hiring a few sustainability consultants, as most companies do today. All CXO need to understand this new agenda in depth to be able to transform the company into the new market situations and to build the culture around a purpose-driven company that future employees and consumers want and engage in. This is not done on the side of traditional business. This requires genuine change of management and often replacement of management (executive and nonexecutive). The largest barrier in the transition is probably lack of knowledge within management.

This means that step one in the transition is a transformation of the educations offered. The educational institutions for young people are struggling but aware of the need for change, some schools have engaged in the material that the Sustainability Development Goals (SGDs) offer, and this is seen successful among the kids in elementary schools and primary schools. The big challenge is vocational schools, high schools, and universities where there is a general lack of interdisciplinary educations teaching the complexity of sustainability and solutions for the transition of companies, markets, and society. Master classes and executive programs are coming out, but the hands-on solutions and understandings for especially the Circular Economy, but also sustainability in general are still scarce.

The universities sit with a lot of science-based knowledge and, with many of the solutions within natural and technical sciences on climate change, and the biodiversity crisis. Not many universities have been able to transfer this into the businesses over the last decades. Neither do many universities offer curriculums on the holistic approach towards the great transition to a Green and Circular Economy or cross-disciplinary research and implementation with the social sciences and the implementation in businesses and in society in general. Stanford University has decided to showcase their research and walk the talk on-site in the large campus around San Francisco, called the living lab or the city in campus. It is worth looking at their work at https://sustainability.stanford.edu/our-impact/living-lab.

Among many analyses, one investigation of the competences needed to support sustainability, and the transition to a Green and Circular Economy was performed in Denmark in 2021, and this resulted in the design of a competence profile, illustrated in Fig. 4.2 (Sloth, 2022).

First, it is important that all the people in an organization build these new competences to be able to implement sustainability. The transition requires a top-down approach, meaning that management must move first and build the framework and the strategy for employees to fulfill. This will not come from a bottom-up approach. Then it is important to remember that knowledge is built with education and training, and competences are built when the trained employees are using their knowledge in their daily work within the organizations.

Figure 4.2 illustrates that it is not only necessary to build new technical skills that meet the complexity of sustainability and the green transition but also to build the personal competences to facilitate the transition and to make the technical competences to play in new ways. Decency, collaboration, communication, innovation,

Fig. 4.2 Competence portrait. The green competences are the technical competences within sustainability and green transition. The blue competences are the personal competences that are necessary to support the technical competences

Knowledge Gap 57

and change management is as important to master as all the new technical skills of ESG, sustainability, and Circular Economy.

> Boards of nonexecutive and executive should care mostly on bringing new skills into corporate management. New skills that meet the future and the sustainable solutions in new ways—if they are not to be overtaken by a bunch of young entrepreneurs.

The lack of ambitious leaders all around, both political leaders and business leaders, has put the planet in a critical state, and now their actions are needed. Starting with them to build their own skills and values to meet the future and to drive their companies towards a fair and sustainable planet.

Picture from a wall in the exhibition of Johnson Space Center, NASA, Houston. https://spacecenter.org/exhibits-and-experiences/. Photo: Gitte Haar

Science Gap

To complete the roadmap of the future leaders, the purpose gap and the knowledge gap are topped by a science gap because there is a huge science gap in companies and business. Research institutions have provided science on the state of the planet and on climate change for decades. Technologic solutions have been developed. Nevertheless, politicians and business leaders are still discussing if climate change is due to human activity and if action is necessary. Most leaders are not aware of the consequences already caused by climate change on the market conditions.

There is a huge gap between what is common knowledge at universities and among intellectuals, and the common knowledge in businesses and society. Somehow all the science-based knowledge created in the last three to four decades have not affected society and businesses enough to change behavior. Science on the state of the planet and sustainability must be an important part of the decision-making processes in politics and in businesses. The EU Green Deal and its strategy on Circular Economy are very much based on science and technological development in understanding the crises of the Anthropocene Age, as well as the climate regulation in the USA (IRA and EO). Still national politicians seem to be stuck in the old linear way of understanding growth and development.

Sustainability and the great transition to the Green and Circular Economy will have difficulties until a scientific approach is taken all the way around.

> **"The world is a dangerous place, not because of those who do evil, but because of those who look on and do nothing."—Albert Einstein**

Apart from the competencies described in the portrait here (Fig. 4.2), leaders face a time where they need to understand and navigate in:

- The full value chain—upstream and downstream and taking not only responsibility for the supply chain but also for the consumption of products and services.
- The geopolitical situation that influences the global economy and the local economies.
- The governance of the company in traceable and transparent ways to ensure the full value chain, financial transactions, and ESG impacts are monitored and managed in accordance with new standards.
- Sustainability is a complex topic that influences everything and is very scientific driven, and not emotionally as many believe.
- History is a strong precaution for culture, and understanding the history of a company and a country in a geopolitical perspective is necessary for corporate leadership.

The leaders of today are meeting a future that is so different from what they grow up in that it will cause resistance, simply because changes of this magnitude can be difficult to comprehend.

"The biggest challenge in meeting new shores is the fear of leaving the old shores."—
André Gide, French Author, and Nobel Literature Prize Winner

Culture and Leadership

Leadership and business culture must change to meet the future and close the gaps described above. Not only to meet a complex future but also to make the companies resilient and agile for the organization to make sustainable decisions. The culture of a company and the country of origin is very important in the way leadership is shaped and performed. A well-known saying is that culture eats strategy for breakfast. Creating a strong corporate culture is difficult and requires strong personalities, strong shared values that go beyond KPIs, and patience. Whereas, the opposite happens very rapidly, destroying corporate culture. Many newly hired CEOs have strong opinions and heavy records on leadership and still give up on managing a corporation due to lack of understanding the corporate culture and the national culture of the origin of the company.

Purpose and new market conditions based on sustainability requires new business culture and new leadership culture. The question is whether the existing executives can take on this transformation on a personal level or if new people and new profiles are required in the corridors of power. As in many cases, some understand and adapt, and others do not. No doubt is it that gender equality and diversity at top level will push organizations in the right direction.

There is a huge difference in managing an American or a Nordic corporation. The Nordic culture described in part II of this book runs very strongly in many Nordic corporations. Even though there is also a difference within the Nordics. Sweden being a country formed by large industries as defense, car production, and other industrial manufacturers holds a different business culture than Denmark or Norway with more primary production (agriculture and fishing) and trade. As mentioned earlier, the Nordics hold some characteristics that have created a Nordic management style admired by many and understood by less. In these years, the Nordic management style has received a lot of attention since the Nordic companies are so strongly represented on the global scene as sustainable. Read more on the Nordics and the business environment here in Chap. 1 and Part II, "SDGs for Transition in a Nordic Perspective."

Strong Nordic business cultures are seen in some of the international corporations that come out of the Nordics as Novo (DK), IKEA (S), Lego (DK), Ericsson (S), Pandora (DK), Skanska (S), Reitan (N), Nokia (F), etc. More on Nordic business and company cases in another book by the same author: Nordic Case Collection for Sustainability.

The historical background for this special Nordic culture may be found in these old democracies. Democracy in the Nordics is strongly supported by an educational system and a democratic understanding that in Denmark took its starting point with a pastor who lived from 1783 to 1872, called N. F. S. Grundtvig. He created a movement to form and educate the people instead of only the nobility. Farmers and rural workers join Formation Schools ("Højskoler") all over the country during the 19th century. Grundtvig created the whole idea that education is the basis for a healthy, democratic, and financially stable society. Implemented as access to free education right up to university level in the Nordics. This is the foundation for an economically and politically stable and equal society.

However, this has not prevented the Nordics today from overconsuming at a level that is now detrimental to the planet. There is now a need to revitalize this formation process created by Grundtvig in the mid-eighteenth century. Today, it is not only farmers who have to be educated. Here, 150 years later, it is the business leaders that need a formation process for the benefit of our shared planet, the green transition, and shared values. There is a general lack of knowledge in the adult population, and they need to understand the interrelations and the long-term effects of the way we live to understand the necessity of changing our way of lives. Here the Sustainable Development Goals (SDGs) may assist. Some business leaders call this business humanism, and there is much inspiration to be gained from Grundtvig's formation theory.

For business leaders revisiting company purpose and the new market conditions, it is still a major challenge in daring to change the business models. The existing business models that have proved profitable, safe, and familiar until now. The new business models in a green and digital economy are untested and based on completely new market conditions requires dramatic organizational changes, changes in business processes, and new agreements with stakeholders.

> We need new sustainable business "Højskoler" (Formation Schools) to create a basic understanding of the need, necessity, and details of the great transition to a Green and Circular Economy. Then we can kick-start the change to a fair and sustainable world.

Nikolai Frederik Severin Grundtvig (1783–1872). Founder of the "Højskole" (formation schools) movement

Grundtvig lived in the time of enlightenment in Denmark and in Europe. He formulated his enlightenment ideas before, during, and after the transition from autocracy to people's democracy that came with the constitution in 1849 in Denmark. He was sceptic towards democracy and believed that democracy is a skill that must be developed among the people for it to succeed. Therefore, he wanted to provide education and formation to adults to bridge or counter the elitist educations evolving in the universities (Latin schools). He believed that the Latin Schools broadened the gap between intellectuals and common people, especially in the rural areas.

His vision was a state-owned Højskole (Formation School) to match the higher educations. The first Højskole was established in 1844 in Rødding in Denmark and created a wave of formation that lasted at least 100 years. Today, the Højskoles are subsidized but typically owned by foundations. The ideals of Grundtvig are also the foundation for Boarding schools for 10th and 11th grade that many youngsters attend voluntarily. Grundtvig's idea of the Formation Schools were a school for life and a place people lived and stayed during their participation (today 1 week to 4 months), and the ideals were:

- All have the right to become a part of community and society, get an education and a formation to be able to participate in society.
- An education or enlightenment of all the people, especially the rural people.
- Education must never stand in the way of human's innate curiosity.
- A free and open conversion is central in human formation and the education of people.
- For young people to gain enough insight into society to be able to take a position on the construction and development of society.
- The people must have a critical and independent voice in relation to the ruling power, whether it is a queen/king, or democratically elected governments.
- Democracy requires freedom of speech, but also adult and enlightened citizen capable of being critical.
- Teaching religion belonged in the church and not in the schools. Nothing must stand in the way for the free and open conversation that must happen at the educational institutions.

The Scandinavian educational system today is formed by Grundtvig's thoughts on information, education, orality, and curiosity. These thoughts and ideals are important to re-implement especially in business and among business leaders, as they become an increasingly significant role in the great transition to a fair and sustainable world.

Company purpose and sustainable leadership is also affecting corporate governance. Corporate governance is defined as "the system by which organizations are directed and controlled." In essence, it is the "rules of an organization's management." Corporate governance is changing and embedding sustainability and greater purpose in shared values with society. A dramatic shift in corporate governance will come with the new due diligence regulation in the EU on extended responsibility in the extended value chain (CSDDD), as described here (Comission, 2023):

> "The EU Commission (2022) has adopted a Directive on corporate sustainability due diligence. The aim is to foster sustainable and responsible corporate behavior and to anchor human rights and environmental considerations in companies' operations and corporate governance. The new rules will ensure that businesses address adverse impacts of their actions, including in their value chains inside and outside Europe."

Globally we must expect stronger regulation and soft law on corporate governance not only due to sustainability and extended responsibility but also due to a shift in the geopolitical situation that will affect how and where companies operate.

> "Life is the greatest adventure."—Hans Christian Andersen (Danish Fairytale Author, 1805–1875)

Time and Ownership Structure

Not only must leadership change to become purpose-driven, and knowledge and competences must be built. Time and ownership structures are other important factors in corporate survival. An economical paradigm-shift to a Fair, Green, and Circular Economy and in a digital and technological world requires innovation, long-term goals, and strategies. The capital structures with equity fund-owned and listed companies have created a tendency towards very short-term goals and short-term employments of executive management. Especially in large corporations. They are hired to harvest large financial gains rapidly, and their salary schemes are built around these short-term financial goals. Maybe all this haste is hurting the survival of business itself.

In the Nordics there is a large share of family-owned, foundation-owned, and customer/member-owned companies. Not only among SMEs but especially among the large global corporations. As, for example, COOP (supermarket) in all Scandinavian countries. Many of the pension funds in the Nordics are customer-owned and managed by the labor organizations (nonexecutive). This is based on old cooperative traditions, which creates a close link between primary producers, customers, and customers. This type of companies seems to have more patience with investments and long-term strategies and thus time to prepare for embedding sustainability and extended responsibility. In the Nordics, the ownership of a significant fraction of global industrial corporations is transferred to charitable foundations, as, for example, Novo Foundation (DK) and Nobel (S). Novo Foundation is one of the world's largest foundations at the size of Melinda & Bill Gates Foundation owning Novo Nordisk (insulin) and Novozymes (enzymes) donating large amount for research, humanitarian, and educational purposes. Nobel Foundation is founded by Alfred Nobel (dynamite), awarding five Nobel Prizes every year and even more money for research and other purposes. Also large global corporations as; Grundfos, Danfoss, Velux, IKEA, and Lego are family owned where the ownership is placed in large foundations that donate significant amounts of money to society.

Time is an important element and necessary in the transition of companies. If the companies and capital do not prepare for this transition and respect the time needed, the overall transition of creating a livable planet will not happen at the speed necessary. Lack of time and capital to the existing corporation may mean that the old business models will be disrupted rapidly. New actors and entrepreneurs not bound by old structures (technologically or financially) or old capital will enter the markets. The transition may simply appear from the bottom and up. Instead of being a transformation of the existing global corporate brands. The trend of bottom-up disruption is already happening due to digitalization giving new access to consumers, as described earlier in this book.

Ownership structure means a lot to the agility of the companies in this great transition. There are many indications that, in contrast to listed corporations, the family-owned, member-owned, and foundation-owned companies dare make long-term decisions and investments to achieve long-term returns. The same does not apply to companies owned by equity funds that typically hold short-term ownerships, where companies must reap large capital returns quickly (5–7 years) with less focus on long-term and sustainable returns.

At the same time, the competences of the capital funds are based on financial analysis of historical market data, historical performance of the companies in a well-known marketplace, and the optimization of hidden asset values rather than development of new long-term business models. There is only little emphasis on navigating a new Green and Circular Economy or holding competences to embed sustainability and responsibility in leadership. Development of new business models requires understanding of the state of the planet and the newly available technologies. Here long-term rather than short-term returns are crucial for companies to survive. As the lack of understanding the ESG (Geopolitics) and SDG in depth is becoming an obstacle of the capital funds and the listed corporations.

The demand for change in companies is not new—it came with industrialization, with the computer technology and again with the Internet. It is interesting to look at the evolution of corporate stock values and turnover over the past few years, which gives a picture of this fast change. A look at the large Nordic corporations delivering according to the *take-make-use-waste* linear business model and their development of stock value over the past few years shows a good example on how fast corporations will face the changing markets. Forced even further low by the corona crisis. Unfortunately, it shows that they do little to adapt to a new sustainable world. Many of these linear businesses try to address the future but do very little in actual changing their business models. Some of these companies have been forced to pull back marketing campaigns on take-back of product, since they were proven so insignificant that it did not comply with the regulations on green claims. It is considered greenwashing when H&M offer return of clothes without ensuring a transparent business model based on reuse and recycling or even information on what is done with the collected clothes.

Many of the major Nordic consumer brands are family-owned, and thus there is no public listed share value to investigate. H&M experienced a massive increase in stock value in the build of a global brand from 2000 to 2015. They saw the highest share price in 2015, but in September 2023 the stock-value is one-third. The share price has been fluctuating since 2017 and is at its lowest in late 2023. This is a stock market where Nordic indexes have been strongly increasing and stabling lately and in market of increased consumption of clothing, strongly controlled by online trading, also offered by H&M.

The transition requires long-term goals and long-term strategies. Short-term targets and shareholders' demand for short-term returns have become a hindrance, and all this urgency is damaging the long-term existence of the companies. Having the time to bring about the fundamental change and meet the great transition of the marketplace will bring a new meaning to leadership and businesses in the future.

There is a need for longer-term investments and long-term strategies rather than what can be harvested within 4–5 years. It takes time to transform companies to a new economy.

The investors behind the private equity funds are often pension funds and other institutional investors who manage funds for a wide range of people. Nevertheless, it does not seem as if pension funds have set the requirements yet to the benefit of their customers/members in the future. Some pension funds, banks, and other financial partners are offering sustainable investments, and the shift towards sustainable investments is being forced by the Sustainable Finance Disclosure Regulation (SFDR) in Europe. Analyses of these sustainable investments showed that many of the sustainable investment portfolios still contain investments in the fossil industry directly or indirectly. It is very difficult to find investment companies with products

> There is a need for longer-term investments and long-term strategies rather than what can be harvested within 4–5 years. It takes time to transform companies to a new economy.

that comply with the ESG requirements on sustainable financing, which confirms the need for legislation on this as greenwashing is also here.

PensionDanmark is an example of a company early to articulate new ESG requirements for non-financial returns and towards responsible management of their investments. They began early to measure gender equality and climate impacts. There is still a long way to go before the financial market, including pension funds, not exclusively focusing on short-term returns even if it is dawns. Fortunately, there is a general movement among pension funds and banks demanded by their customers to impose sustainability requirements on their investments and recently supported by the EU legislation on Sustainable Finance Disclosure Requirements (SFDR).

SME Versus Corporates

It often turns out that small- and medium-sized companies are more adaptable than large corporates. It is well-known that innovation disappears in the silos needed for scale. Those large corporations are deadlocked by the channels they have built towards the market. The large corporations have proven experts in streamlining and optimizing their value chains. In doing so, the power of innovation and agility has disappeared. Here, SMEs and the owner-managed companies are given new opportunities with sustainability. The decision-making processes in SMEs are shorter, and the owners typically sit at the helm and dare to make long-term investments.

70–80% of the European business are SMEs, defined as revenue below €50 million and less than 250 employees. The labor market is split differently as a larger portion of labor is employed in corporates. It is important to find the connection between the agility of SMEs and the strong muscles of implementation in the corporates. Initiatives to do so are in place but the landscape of business is still fragmented. Today the corporates are mainly a struggle for the SMEs and a barrier for innovation.

> **Many corporates**
> - Push long payment terms also against the SMEs.
> - Set demands on quality and delivery that excludes many innovative SMEs even when, offering new sustainable solutions and products.
> - Hold long-term vendor agreements that slows down the transition through new supplier.
> - Do not leave room for innovation in new product solutions even on a test basis,
> - Are not able to make the final decisions on innovative and sustainable solutions because many heads in the corporates need to be involved, and decision across silos is almost impossible.
> - Are not loyal to inventors, and examples of good innovative solutions and prototypes that have been stolen and handed over to existing suppliers to finalize are many.

These barriers must be overcomed, and the SMEs must be met in new ways, so corporates can embrace their agility and innovation for both parties to transform and survive the new market conditions. Corporate innovation labs are emerging within many industries but are mainly targeted at entrepreneurs and startups, not the existing SMEs and their innovative solutions.

References

Comission, E. (2023). *Corporate sustainable due diligence directive – DRAFT.* Retrieved from Comisssion Europa EU: https://commission.europa.eu/business-economy-euro/doing-business-eu/corporate-sustainability-due-diligence_en

Haar, G. (2024a). Chapter 6: EU regulation to a green economy. In G. Haar (Ed.), *The Great transition to a green and circular economy.* Springer.

Haar, G. (2024b). *The Great transition to a green and circular economy.* Springer.

Sloth, G. H. (2022). *AFRAPPORTERING – Bæredygtighed og Grøn Omstilling (Danish).* Uddannelsesnævnet HAKL.

Chapter 5
New Economics Need New ESG Data

Companies will experience a demand for new types of data to meet legislation on the new economics and support the great transition to a Green and Circular Economy. A transition of society and businesses calls for transparency and traceability in all the processes and impacts along the long value chain.

> **The data to support transparency and traceability are on various levels, as follows:**
> (a) ESG data on national level
> (b) ESG data on company level
> (c) ES data on product level and several non-financial additional data
> (d) Data on the exploitation and impacts on externalities collected from the above data
>
>   E = environmental, S = social, G = governance

EU Legislation on Non-financial Data

Definitions and legislative requirements on non-financial data on companies and products are evolving rapidly these years. EU is establishing a framework to facilitate sustainable investments, Sustainable Finance Disclosure Regulation (SFDR). (See https://eur-lex.europa.eu/legal-content/EN/TXT/?uri=CELEX:32020R0852.) This regulation establishes the criteria for determining whether an economic activity qualifies as environmentally sustainable for the purposes of establishing the degree to which an investment is environmentally sustainable. This regulation applies to (a) measures that set out requirements for financial market participants or issuers in respect of financial products or corporate bonds that are made available as environmentally sustainable; (b) financial market participants that make available financial products; and (c) undertakings which are subject to the obligation to publish a non-financial statement or a consolidated non-financial statement. This regulation will make sustainability and sustainable investments more transparent and will enforce new types of non-financial business data. More on this in another book by the same author (Haar, Chap. 6: EU regulation to a green economy, 2024).

The EU corporate sustainable reporting directives (CSRD) have been introduced, and all other companies than the financial partners are subject to this new reporting regulation over the years to come and for listed companies as of 2024. Product environmental footprint (PEF) and Digital Product Passport (DPP) are being introduced with the sustainable product regulation (SPR) and new eco-design criteria. All this because of Roadmap of the EU Green Deal and the implementation of the Action Plan of Circular Economy and the extended producer responsibility (EPR) on products to make the European economy independent on import of virgin resources and the impacts on the environment and climate. Additionally extensive legislation on substantiating sustainability claims to counter greenwashing. Most important is that the responsibility of businesses owners and managers is being extended to the full value chain (scope 1 + 2 + 3) for products put on the EU market and companies operating in the EU, by the Corporate Sustainability Due Diligence Directive (CSDDR) adopted in 2024. This is the most extensive and rapidly introduced legislation put on companies long seen.

Additionally, EU is introducing new legislation on a personal digital ID for every EU citizen to be able to protect citizen personal data and thereby extending the GDPR regulation. All this new legislation together with the Corporate Sustainability Due Diligence Directive put on companies to monitor impacts in the full value chain will completely change the EU market and business landscape. All this and much more puts the companies and the manufacturers in- and outside EU in the center of collecting, registering, monitoring, and reporting on company level and on product levels. More on EU legislation in another book by the same author (Haar, Chap. 6: EU regulation to a green economy, 2024).

These are large amounts of new data that companies must handle as of now. KPMG just (2023) published an investigation stating that 3/4 of all companies, globally, are not able to identify or collect the data necessary to meet the ESG disclosure requirements (KPMG Research, 2023). Implementation time has been extended on the mandatory data, but the framework, the taxonomy, and the Sustainability (ESG) standards are put in place now.

> Sustainable impacts are based on the SDGs and defined as ESG—environmental, social, and governance—and described in detail in the EU sustainability standards, 2021.

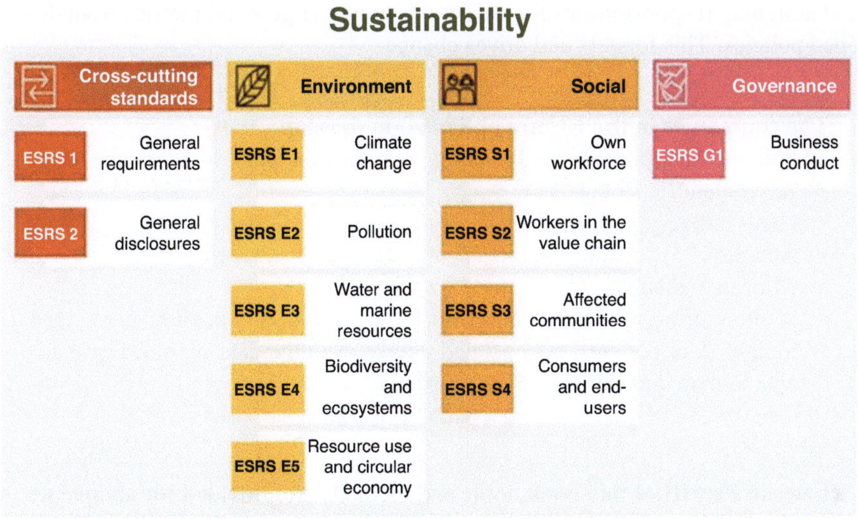

Sustainability

Cross-cutting standards		Environment		Social		Governance	
ESRS 1	General requirements	ESRS E1	Climate change	ESRS S1	Own workforce	ESRS G1	Business conduct
ESRS 2	General disclosures	ESRS E2	Pollution	ESRS S2	Workers in the value chain		
		ESRS E3	Water and marine resources	ESRS S3	Affected communities		
		ESRS E4	Biodiversity and ecosystems	ESRS S4	Consumers and end-users		
		ESRS E5	Resource use and circular economy				

Fig. 5.1 EU sustainability standards. The EU sustainability standards are part of the disclosure requirements put in place in the EU. They are described in the European Sustainability Reporting Standards (ESRS) that links to Sustainable Finance Disclosure Regulation (SFDR) and Corporate Sustainability Reporting Directive (CSRD) that all companies operating in the EU will be subject to within the next 4 to 5 years

The EU sustainability standards on ESG is illustrated in Fig. 5.1, illustrating the reporting requirements of companies (SFDR and CSRD). This is further described in detail by EFRAG based on the European Sustainability Reporting Standards (ESRS) (https://www.efrag.org/lab6?AspxAutoDetectCookieSupport=1).

The framework and detailed requirements defined by EFRAG are very substantial and well-prepared and support the science on the state of the planet and the actions needed to create a fair and sustainable living.

The 12 guidelines include 2 general requirements and disclosures, 5 new environmental reporting standards, 4 social reporting standards, and 1 business conduct reporting standard, all linked to the EU taxonomy. First, it is a set of new data requirements and definitions of KPIs, but most importantly are the general requirements imposing new requirements on companies. Not only are companies to report annually, but they must also implement a ESG governance structure in the company and describe how issues on ESG impacts are monitored in the organization, and how they ensure involvement through the organization from top management in handling of the ESG issues. It includes an educational obligation of all relevant employees and assigning responsibilities for the actions to met targets and metrics from developed policies. This forsters and drives change.

> **The companies in the EU are obligated to develop:**
> - Policies and procedures including a transition plan to meet the Paris Agreement
> - Goals and targets
> - Actions
> - Allocate resources
> - A monitoring framwork describing the short-term, medium-term, and long-term in policies, actions, and targets. And the data-monitoring available to management on a timely basis.

Later in Part III of this book, tools and methods are provided for companies to comply with all the non-financial data (ESG) requirements. Tools are available in digital forms in various reporting tools.

It is important to understand that all the new ESG data are not just for complying to disclosure regulation. They are necessary to drive the change to a fair and sustainable world and to support the transition to a Green and Circular Economy. Many companies are approaching this new EU legislation to be able to provide an audited annual report, but the requirements also request an integrated ESG governance structure and timely monitoring of policies, targets, and action plans. Soon the requirements on action and strategy will also be tightened from the financial partners. For years, stock exchanges and credit institutions have put ESG data requirements on public-listed companies, and EU has had non-financial data requirements (NFR) and, for many large corporations reporting on ESG, is not new.

Now the legislation also requires action, and then data becomes the basis for companies to manage the transition in the short and in the long run.

> The keys in the EU ESG reporting are the extended value chain covering scope 1 + 2 + 3 and the life cycle analysis (LCA) of products.

Documentation of the full value chain based on LCA is the basis for communication about company impacts and making green claims.

More and more companies commit themselves to various standards and associations to find inspiration and support in monitoring and driving change, as:

- UN Global Compact
- Science Based Targets
- B-Corp
- Cradle to Cradle
- Others

Many of the abovementioned semi-private standards are now being criticized by scientists for not meeting the demands of life cycle analysis (LCA) stated in legislation but may still be for inspiration. Thus, it is important to ensure that companies can comply and align ESG data and KPIs in the full value chain (scope 1 + 2 + 3) with the SFRD/CSRD requirements, as well as with the increasing demand for product data. In the EU, ESG data in the annual reports are subject to external verification by auditors, today often the same auditor performing the financial audit, even though they are struggling to understand and assess the ESG report.

International accounting standards are being developed for sustainability data and ESG that will become international standards across regions. IASB (International Accounting Standards Board) that is responsible for IFRS (International Financial Reporting Standards) has appointed an International Sustainability Standards Board that is preparing international standards for sustainable data. The international independent standards organization, GRI, helps businesses, governments, and other organizations understand and communicate their impacts on issues such as climate change, human rights, and corruption. This has existed for some time and has been voluntary standards that companies could adapt to. Having international accounting standards for ESG data as known from the financial data is important to assess and compare companies on their footprint on the planet.

One example of scoring companies on the ESG performance is Refinitiv ESG Scores, which provides a tool for scoring companies, based on data from 9000 exchanges, and has comparative data on corporate performance (Finansdata, 2019). The tool is available and can give insight into various companies and parameters of ESG for comparison.

The various stock exchanges and investment banks around the world have developed sustainability indexes for investors, such as FTSE, MSCI, and others. They rely on self-developed standards and systems in the field of ESG. Largely, all the abovementioned standards, frameworks, and tools are targeted at large corporations. Especially the international corporations make use of these tools. The tools are very comprehensive and expensive to work with, and this is typically outside the capacity of SMEs. With the new legislated standards on reporting, this will become more uniform across borders and organizations that require these data. *A Corporate Accounting and Reporting Standard* published by World Resources Institute of World Business Council for Sustainable Development are also part of the development of international disclosure standards that will emerge and unify the years to come.

ESG Data Ecosystem

For years many companies have announced ambitious climate neutrality targets and other sustainability goals, and now it is required for them to document their company impacts and the impacts of their products. Corporate sustainability targets require transparency and traceability in the full value chain of products and

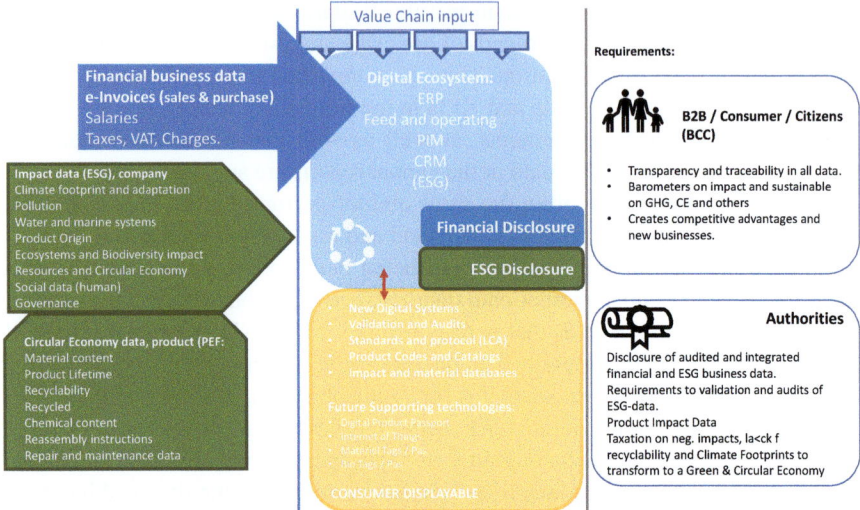

Fig. 5.2 Ecosystem for financial and ESG data. The figure shows in dark and light blue the systems we know today for handling financial data. The green figures show the ESG data that companies must report on as of now, and the white figures on the right side of the illustration show where the requirements for new data and transparency come from. At the bottom in the yellow box are the indicated systems, catalogues, and standards needed to support a uniform and comparable handling and validation of ESG data

services—from production, transport, consumption, and end-of-life to life cycle analysis (LCA), fully in line with legislation.

Figure 5.2 shows an example of an ecosystem for new ESG data. The illustration has been prepared for Nordic Smart Government (NSG), which works on the development of an ecosystem for financial and non-financial data to support SMEs in the Nordic region. This project is followed closely by the EU and could be one of the first examples of a comprehensive framework to support the handling of both financial and non-financial data (Haar, Memorandum. Non-financial business data for SMEs in the framework of Nordic Smart Government, 2019).

Especially for SMEs, managing and capturing these new types of business data is a challenge. It is important that governments and institutions enable SMEs to display sustainability data on their products, giving them a competitive advantage that supports their transition and gives them access to new export markets. There is a need to support and alleviate the burden of the SMEs even if they are already met by non-financial data requirements from the customers when being larger corporations.

The overall goal of a transition to a Green and Circular Economy is to preserve as much economic value in the products and materials as possible, creating growth and jobs without continuously exploiting nature. We need to make use of the materials and resources available within the EU, as access to virgin materials becomes harder and the prices of raw materials rise. The Circular Economy creates growth and new market potentials when companies harvest the values from what is today thrown away as waste or combusted for energy production. In this way, we can maintain the high level of prosperity in regions with no or little raw materials and in the situation of global scarcity of raw materials. Implementation of Circular Economy and sustainable products requires that companies, financial partners, and authorities have access to standardized data. The EU's new sustainable products initiative (SBI), the PEF (product environmental footprint), and the DPP envisage measuring products according to life cycle methods (LCA) and thereby allowing benchmarking between products (6). This requires standardized methods for performing the life cycle assessment (LCA) of the products, and this again requires standardized catalogues and databases of the impacts from various products and materials and from the various processes throughout the value chain. This will certainly lead to new product codes and new product and material databases containing technical data on the impact of products on several sustainability parameters. The Digital Product Passport (DPP) is the carrier of all this data. The sustainability parameters must be linked to the basic elements of the ESG impact data fostering the Green and Circular Economy.

Monitoring not only climate impacts and impacts on ecosystems, chemicals, and pollution but also data on product and material origin and the actual and potential reuse and recycling is necessary for companies to document that their products are made for a clean and Circular Economy from recycled and recyclable materials. The standard protocols for ESG data will be developed by the legislators and the universities in cooperation with the industries and the industry associations that have knowledge of materials and material flows.

Not only legislators and financial partners are demanding ESG data from companies, but consumers are also to a larger extent demanding transparency and traceability from company and product impacts. The material and product databases (or catalogues) will form the basis for companies' assessment of their impacts throughout the value chain and will be updated and verified by external parties. Exactly, as currently known from the financial auditors. The big challenge in ESG data is on access and validity of scope 3 and will be very interesting to follow the evolvement of the standardization of these data. Another challenge is how product data are transformed in the the ESG data that are to be reported on company level in scope 1+2 and 3, here the LCA databases and data-enriched DPP will be the key, still lacking the systems to do so.

Examples of ESG Data Supporting Sustainable Transition

EU has a strong focus on the construction industry due to this industry's large climate footprint and waste production. The Netherlands is a showcase for the Circular Economy and launched a project to register the entire standing building stock at material level to be able to supply the future constructions with reused and recyclable building materials to meet industry reuse/recyclability targets of 50% by 2030 and 100% by 2050. The Netherlands are almost finished with this registration, and Denmark has started as of 2022, as well as developing the blueprint and shared framework for a material passport. This demonstrates how important non-financial data are on product level.

At the end of 2020, selected countries and companies in the EU adopted a new Plastics Pact, which includes a design guide to and by the plastics industry on how to make plastic recyclable. Here are requirements for categorized, uniform plastic polymers, labeling of the products and traceability of the plastic through the recycling system. With the new DPP and PEF, requirements will be added for accounting for impacts on climate, biodiversity, and other matters to be used in benchmarking the environmental footprint of products. The European Plastics Pact contains guidelines and recommendations for a new circular design manual for the entire industry and has been developed in collaboration with retailers to meet consumer demands for less single-use plastic, recycling plastic, minimizing greenhouse gas emissions, and avoiding pollution of nature and especially in the marine environment. This is strongly supported by the extended producer responsibility (EPR) that is put on packing material as of 2025, as 30% of all household waste is packing material to a large extent from plastic. The EPR on products in EU requires designated and concrete data, and the transition must be managed and driven by data.

Upgrading industry requirements, data requirements, and design manuals will be introduced in all industries and for all types of materials by 2025, where also the first DPP will be available for products. A lot of the EU legislation within Circular Economy and sustainable product initiative will have to be implemented by 2025.

Consumer Demands and Labeling

When demanding Value Chain Transparency on several types of sustainability data, consumers now turn away from the jungle of private and semi-private ecolabels. The private and semi-private ecolabels cover different impacts and not always according to the principles of LCA. They have become too complex to understand for both procurers and consumers. Consumers have realized that the different ecolabels often cover only few sustainability parameters and often not cover the full value chain. It is difficult to find ecolabels complying the new ESG standards. The textile and leather label ÖKOTEX and the food label "Organic" are examples of an ecolabels that do not include climate impacts, biodiversity, circular impacts, and so on. Experience over the last few years has shown that consumers and procurers demand more objective criteria for assessing products, thus increasing the need for independent consumer barometers and more standardized information, as seen with the product environmental footprint (PEF).

A small example of this was when IRMA (member owned sustainable food retailer under COOP in Denmark) asked their members (customers) about sustainability, packing material, and future sustainable foods back in 2016. The outcome was clear. The customers want an independent barometer on product footprint validated by the retailer. IRMA's customers were first movers in food sustainability, quality, and innovative products. Today IRMA has closed because management chose a corporate strategy to standardize and unify the sourcing of products for the whole COOP group which led to less sustainable and specialized food products in IRMA for the trendsetters that were loyal to IRMA. Additionally, a lot of IRMA private label products were developed by COOP, and customers lost track on local producers and products that is important to be able to assess food. The case study of IRMA is an example of spreadsheet management away from understanding customer needs and away from a sustainable passion for food that killed an old, high-level retail brand in Denmark in very few years.

The overall conclusion is that ESG data on company level and product level is the pavement to drive the transition to sustainability. Good intensions are good, but facts on impacts, traceability, and transparency are necessary to keep greenwashing from the door. To make customers and procures pay a premium for sustainable products and enter strategic collabs with companies, it must be much clearer what and who are genuine sustainable and who is using illegal marketing. Companies are struggling with the new ESG disclosure regulation, but this is the necessary first step to facilitate the change to a fair and sustainable planet that we are in the need of to support humanity.

The yearly costs of ESG data collection, monitoring, and reporting are estimated by EU Commission to 3.6 billion € in 2024 where 50,000 companies are subject to reporting. This number will only increase as more and more companies become subject to reporting and when the standardized ecolabeling of products (PEF and DPP) are introduced as of 2025.

References

Finansdata, E. R. (2019). *Top Ten Companines by Market Cap over 20 years*. https://ercouncil.org/2019/top-ten-com-panies-by-market-cap-over-20-years: ercounsil.org.

Haar, G. (2019). *Memorandum. NON-FINANCIAL BUSINESS DATA FOR SMEs IN THE FRAMEWORK of Nordic Smart Government*. https://nordicsmartgovernment.org/.: Erhvervsstyrelsen.

Haar, G. (2024). Chapter 6: EU regulation to a green economy. In G. Haar (Ed.), *The Great transition to a green and circular economy*. Springer.

KPMG Research. (2023). *Regulatory deadlines loom and only 25% of companies feel ready to have ESG data independently assured. Road to trust: KPMG ESG Assurance Maturity Index*. https://kpmg.com/xx/en/home/media/press-releases/2023/09/kpmg-esg-assurance-maturity-index.html: KPMG.

Chapter 6
Organizing the Company for the Green and Circular Economy

Not only the external conditions need to change, so does the way that companies are organized. This is necessary to meet the autonomous challenges and the new market conditions due to implementing sustainability and the transition to a Green and Circular Economy.

Circular Economy

Here follows a short introduction to the Circular Economy in a company perspective, as companies are the driving wheels in this transition. Read more on Circular Economy in another book of the same author (Haar, Chap. 9: Transition to a Circular Economy, 2024b).

The Circular Economy will require a much closer cooperation with customers, vendors, and the recycling industry. Resulting in new business models in the value chain and new collaborations to handle the used products and materials to create the future material banks. Then companies will be able to fully account for their environmental footprints and operate according to the principles of the Circular Economy based on recycling and reuse.

The production of waste in a linear economy has caused and is still causing tremendous environmental problems all over the globe. Especially in the regions where the products are produced and less in the regions where they are consumed. Europe has been exporting waste to third world countries in irresponsible ways. This goes for plastic, electronics, textiles, and other types of products. The linear consumption has a huge impact on climate and nature, and it is estimated that up to 45% of GHG emissions relates to consumption of products, mainly produced in Asia and consumed elsewhere typically in Europe or North America (Haar, The Great Transition to a Green and Circular Economy, 2024c).

In the future, the global economy will be limited by access to material resources, as it was in the years before and during the World Wars in the last century. Today, resource dependency is due to overpopulation and overconsumption in linear ways causing drag on resources and nature. Earlier as in the beginning of the last century, resource dependence was due to lack of technology to extract raw materials and difficulty in transporting these. This changed with technological development during the industrialization.

> **It becomes crucial for societies and companies to:**
> - Ensure access to raw materials
> - Be able to predict prices for their input materials
> - Operate sustainably without negative impacts on nature and the climate, and with full traceability in the value chain
> - Document sustainability in transparent, clean material loops

The EU has Circular Economy as one of the important pillars in the EU Green Deal, as the roadmap to meeting the SDGs. The transition of European companies to Circular Economy within the next 5–8 years is a reality (Haar, Chap. 6: EU regulation to a green economy, 2024a) and will strengthen the European economy. This will change the market situation and the competition between countries and between continents. See more on EU legislation in another book of the same author (Haar, Chap. 6: EU regulation to a green economy, 2024a).

An important change companies will take on is a different view on the customers. In a Circular Economy, consumers will become suppliers of used products and materials. This way companies can directly and indirectly ensure access to materials through take-back systems. Involving recyclers, resource banks, and other strategic partners. Products must be designed for longer life, for repair and maintenance, and for disassembly to recycle the materials according to the resource hierarchy or value pyramid (Haar, Chap. 9: Transition to a Circular Economy, 2024b) as illustrated in Fig. 6.1.

The circular value chain covers the extended value chain, the new way of organizing material flows and loops, and is described thoroughly in another book by the same author (Haar, The Great Transition to a Green and Circular Economy, 2024c). The circular value chain illustrated as an infinity sign in Fig. 6.2 is showing how the economies value chains and companies become independent on virgin inputs and the generation of waste by closing the loops of products and materials. Overproduction is a significant problem in many industries. In the fast fashion industry it is estimated that 30–50% of all clothes are not purchased by the consumer but descared and destroyed. The building industry estimate around 20% of all building materials are never installed but discarded already at the building site due to cut-off,

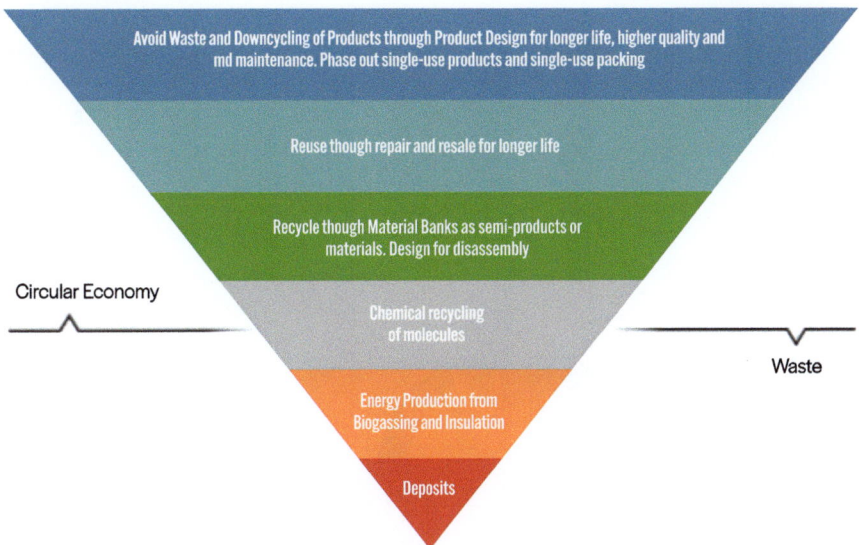

Fig. 6.1 Resource hierarchy. New eco-design criteria and extended producer responsibilities (EPR) will enforce a Circular Economy in EU through legislation based on the overall principles of value/resources hierarchy illustrated here

unfinished batches, or mistakes in measuring or fitting the materials. In the production of food it is estmated that 30% is wasted in the full value chain. All this overproduction is due to long, linear value chains and industrial seriel manufacturing set-up. This is a meaningless and unnecessary waste of resources and values that most be avoided at the top of this Resource Hierarchy.

Europe and Asia are already facing a lack of important raw materials, as rare soil minerals and the metals used to produce clean tech and ICT, all important in the green transition. Companies are experiencing increasing prices, and especially in Europe and Asia, scarcity will become vital due to high population density and dependency on imported raw materials. Here the focus on the Circular Economy is rising. Also, North America has experienced scarcity due and dependencies on Asia during COVID and after with increasing consumer prices, etc. The transition to a Green and Circular Economy is becoming global to stabilize economies and make the regions more self-supporting. EU is introducing an act on Critical Raw Materials including recycling of materials to meet the transition and the future demands within EU.

80   6 Organizing the Company for the Green and Circular Economy

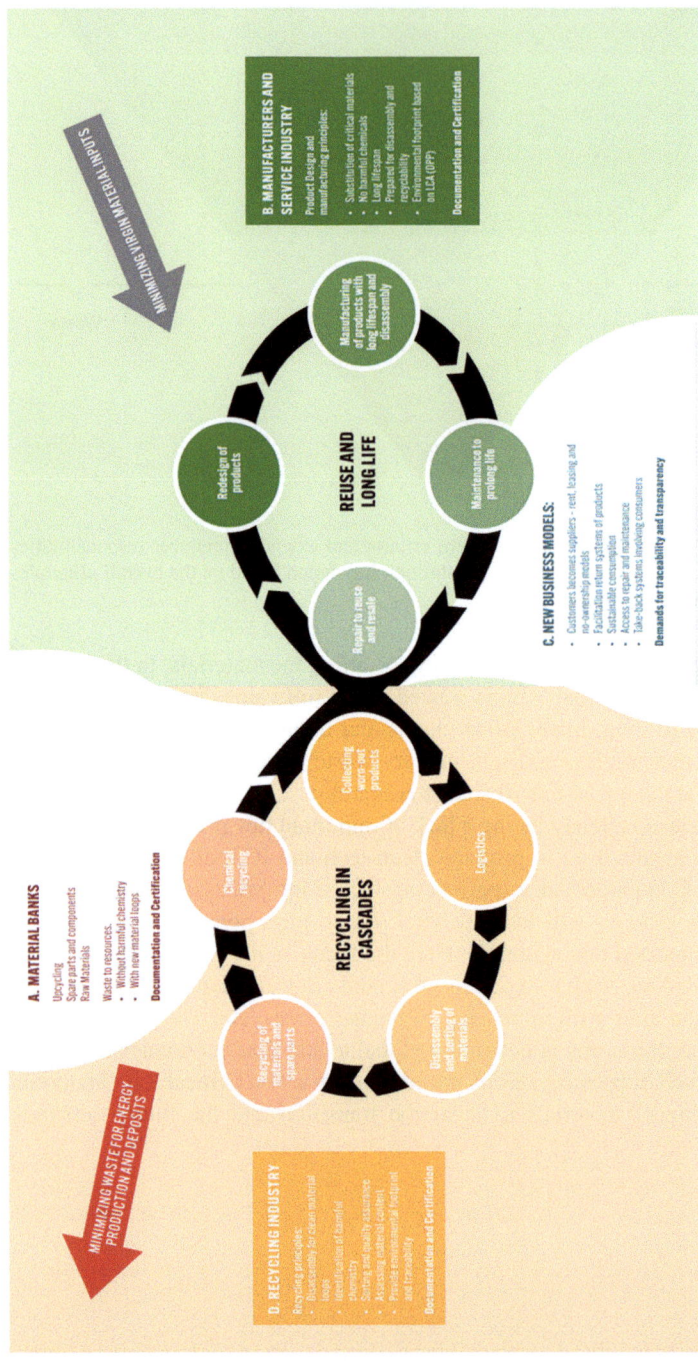

Fig. 6.2 The circular value chain. The circular economy is to create independency on virgin inputs and to abolish the waste creation. This requires new value chains based on long lifespan of products with use, reuse, maintenance, repair, and refurbishment (right green side). And recycling of materials through logistics disassembly and new materials banks (left yellow side). All this is driven by many new business models in the full value chain and of the principles of Circular Economy that must be embedded already when products are designed and redesigned

The Circular Organization Model

Back in the 1990s, M. Porter's value chain created the fundamental understanding of our businesses, their organizations, and their activities. Porter's value chain illustrated in Fig. 6.3 has been at the core of understanding businesses and the way they are organized for leaders for decades (Porter, 1985).

Porter's value chain is developed in and for a linear economy and was the way to describe a business and the organization back in the 1990s (Porter, 1985). This way of approaching business activities and streamlining the processes is still the way most companies are organized. Porter's value chain has become a religion in the business world. Most interestingly, it was called a value chain, but only describes the activities within the company and the flow of products and information through the company. This may be one of the keys to the latest decades' one-sided focus on developing the linear value chains. Not including the full upstream or downstream of a products supply chain, thereby not assessing the full value chain. This clearly illustrates how academia and business leaders have focused for decades in developing the linear economy. Now it is time for change to meet the needs for sustainable businesses. The first step is to unfold the full product value (supply) chain and understand all the ESG impacts in this extended value chain. For more details on value chain assessment and ESG, see other books by the same author on the great transition to a Green and Circular Economy (Haar, The Great Transition to a Green and Circular Economy, 2024c).

> Porter's value chain is a description of organizing a company in the linear economy, while this new circular organizational model describes the company's organization in a Circular Economy.

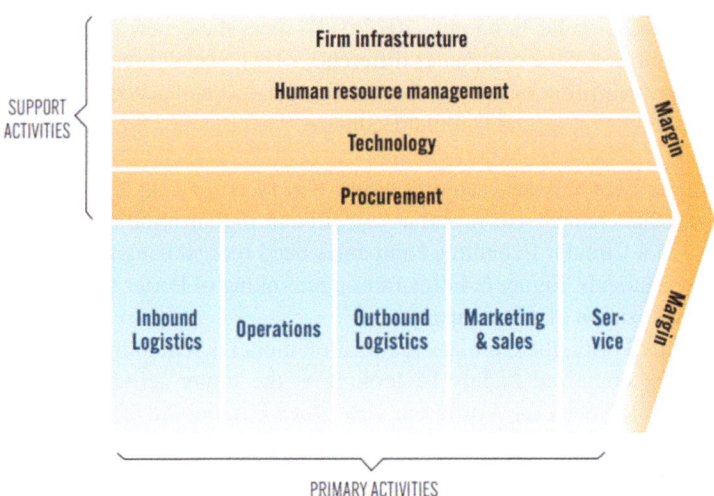

Fig. 6.3 M. Porter's value chain. Porter's value chain has been important for understanding business for decades and is a way of organizing a company in a linear economy

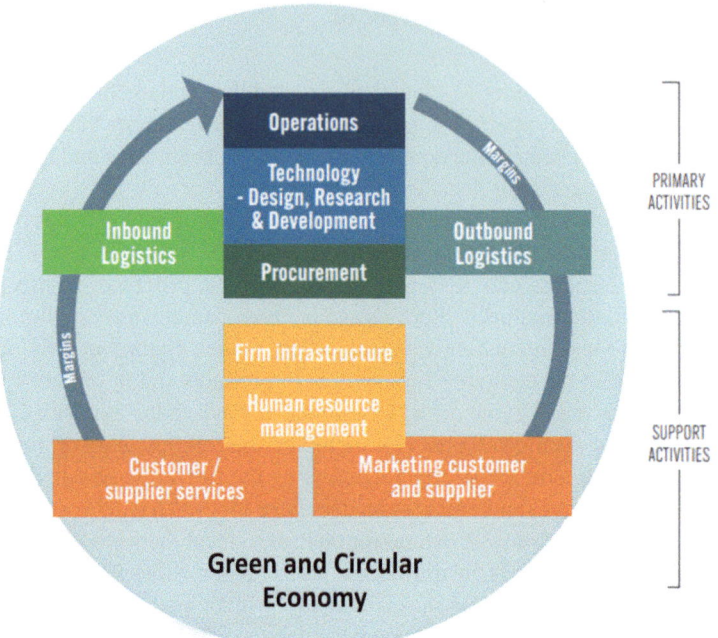

Fig. 6.4 The circular organization model. Reorganizing activities and support functions in companies are necessary in a company operating in a Circular Economy

Companies adapting to the new Green and Circular Economy must develop business models and new product based on sustainability and resource efficiency. The declining access to virgin raw materials and the increasing prices on materials are currently forcing business leaders in Europe and in Asia to develop new business models based on the principles of the Circular Economy (Haar, The Great Transition to a Green and Circular Economy, 2024c). To meet these new requirements, it is necessary to organize the companies differently. This chapter presents a reorganization and reprioritization of the activities and support functions described by Porter to meet the future Green and Circular Economy. This is called the circular organization model.

A Circular Economy also requires a new way of organizing and operating companies, and this chapter will reveal a new circular organization model inspired by M. Porter. In a Circular Economy, companies need reorganizing to be able to meet all the new demands. Figure 6.4 illustrates a rethinking of Porter's value chain for a circular organization of the company.

Inspired by Porter, the circular organization model is rethinking the way a company must be organized mainly be looking at the future activities in a Circular Economy and the need for a different way of prioritizing the primary and support activities, but also on how to organize the company processes to meet the future. The main difference between Porter's value chain and this new circular organizational model is a shift between primary activities and support activities to support the production of sustainable products and most importantly to navigate the supply chains that the company is an integrated part of. Procurement becomes managing sustainable procurement crucial to control the supply chains and the loops of raw materials and sub-products and where and how these are manufactured, transported, maintained, and recycled.

The departments in the company will interact in new ways, creating new relationships and new dependencies within the company. Companies must change their organization, but above all they must change the tasks to be carried out by the various departments, as:

- Primary activities, as sales and marketing, become support activities.
- Procurement and technology, which were support activities, will become primary activities in the new circular organization of companies to ensure sustainable, circular, and traceable products.
- Design, development, and production of new products and new business models will be the most important in a Circular Economy.

"Be the change that you wish to see in the world."—Mahatma Gandhi

The major changes are:
A. Production together with technology, design, and development of circular products will be the company's most important activities in the new green economy.
B. Customer and product service will be a primary activity to increase the durability of the products, and the company must ensure maintenance, reuse, and recycling of the products.
C. Sustainable procurement strategy and policies will become a primary activity as an important gatekeeper of all corporate procurement. Inbound and outbound logistics are moving closer to purchasing.
D. Marketing and sales as support activities, and both customers and suppliers must be serviced by people holding strong circular and sustainability competencies.

This is unfolded here:

(A) Operation together with technology, design, and development of circular products will be the company's most important activities in the new green economy.
In an economy based on scarce resource, the products and the documentation of their sustainable life cycle are crucial for the company to be able sell the products on the market. The new EU Green Deal is creating a new marketplace for sustainable products and requirements on how products are manufactured. Their sustainable value chains become an important competitive parameter and a license-to-play

on the European green marketplace. Technology and operations including design, R&D, ESG, and material understanding will be the most important activities in the circular organization. Technology was a supporting activity in Porter's value chain.

Technology will play a completely new role in future companies—due to the fast development of digitalization and automatization (Industry 4.0). Knowledge about the newest technologies and the products is also essential in developing new products for a Circular Economy. An understanding of the impacts from our products and materials is necessary to a much larger extent than companies have today. Technology also offers new business models due to an easier connection directly from producer to customer through web and digital platforms. Therefore, technology becomes a primary activity.

In the future, we will see a need for a much tighter cooperation between operations, service, and technology, including design and R&D. Companies must design and develop products that disassemble, have a longer lifetime, and can be included in a take back system or are recyclable as raw materials or semi-products.

Technology and the circular transition promote each other internally and externally in the company. Management must understand both these topics and their interrelation to make companies survive. Knowledge on the latest technologies, products, and their materials are also important in the development of new products for a Circular Economy.

Digital technologies can create traceability and transparency at product level, which is a necessary to document the impacts of products and materials in the full value chain. This will be required to a much larger extent than companies are used to today. The new product environmental footprints (PEF) that EU introduces with new Sustainable Product Regulation (SPR) and new eco-design criteria will require assessment in the full product lifecycle and data maintained by the new Digital Product Passport (DPP) that every product will carry in the future.

Digitalization and new technologies are also a part of the Circular Economy and may also result in new business models engaging customers and suppliers in new ways and causing a more direct connection between manufacturer and customer via the digital platforms, as seen already.

(B) Customer and product service will be a primary activity to increase the durability of the products, and the company must ensure maintenance, reuse, and recycling of the products.

Products in a Circular Economy must last longer and be designed for repair and reuse according to the resource hierarchy in Fig. 6.1. This places new demands on the service functions of the company. They will move closer to both sales, operation, and technology to able to design, manufacture, and service through the life cycle of a product. Today, reuse, recycling, and retrofitting of products are often done elsewhere than in the original manufacturing company, but it does not have to continue like this. In the future, producers will be subject to extended producer responsibility (EPR) which also covers the longevity and durability of the products promoting many product cycles and new material loops. New opportunities and new business models based on reuse of products and recycling of materials to extended product lifespan through repair and maintenance, for the purpose of containing as much value as possible. In contradiction to the linear economy, where the "goal" is to produce for waste in a flow of *take-make-use-waste*. Good quality and well-maintained products can be included in the manufacturer's own recycling and reproduction processes. Then cash flows will shift from a direct income from sales to business models driven by the Circular Economy and the maintenance of value and resource in the new economy.

In a Circular Economy, the customer often becomes the tenant of the product and thus also a kind of supplier when the products are returned to the manufacturer. The circular loops of used products and materials ensure that the company has materials available for new redesigned and refurbished products or has supplies of semi-finished products for future production. With the increasing reliance on used and recycled products and materials in the supply chain, with customers as users rather than as owners, the return systems are building a new type of relationship between provider and customer. This can create brand and product loyalty in new ways for businesses (Haar, Chap. 6: EU regulation to a green economy, 2024). The new business models that forster the Circular Economy will also change financial position of the assets as the manufacturer may become owners of the product and the user the tenant.

(C) Sustainable procurement strategy and policies will become a primary activity as an important gatekeeper of all corporate procurement. Inbound and outbound logistics are moving closer to purchasing.

Genuine sustainability means far more documentation, transparency, and labeling of products and businesses. Understanding the footprints of purchased products and materials is necessary. Thus, sustainable procurement becomes a primary activity together with inbound and outbound logistics to be able to operate a sustainable business. Traceability of contents and chemical substances will become much more important in the future, and the ability to trace recycled materials as well as document recyclability will be an important part of operations and procurement to a much higher extent than today.

Inbound and outbound logistics will no longer be two separate activities. These two logistic activities will become related because the returning of products and receival of recycled materials is as important as the outbound logistics in the circular business model. They will largely become combined activities because the take-back of products or the receiving of recycled materials will be as important as shipping the of new products. In the future this is tightly connected to sustainable procurement, as it becomes equally important to service vendors as customers in a connected and traceable value chain.

In the future, the freight partners will cover both future in- and outgoing shipments for transport to become efficient and sustainable. Read more on sustainable transport in another book by the same author (Haar, Chap. 10: Transition to Sustainable individual and public Transport, 2024). Company sales will be conditioned by the return of products or materials via various kinds of deposits and refund systems, recycling logistics, or otherwise. Not necessarily a direct 1:1 return system, but a system facilitated by strategic partners from the recycling or freight industry. In practical terms, there may still be physical distance in the handling of incoming and outgoing materials and goods in the factory, but as a business activity, they will be administratively and contractually closely linked, due to the dependencies in the business models and sales contracts.

(D) Marketing and sales as support activities, and both customers and suppliers must be serviced by people holding strong circular and sustainability competencies.

Marketing and sales will (again) become support activities as they were in the early days of industrialization. In the latest decades, marketing and sales have been the primary activity of a company, especially in the consuming Western countries. Marketing and sales have moved away from each other. Branding and creating a lifestyle storytelling about any product to capture customer attention becomes more and more important. Sales is to a large extent a follow-up on marketing activities with the purpose to hand as many goods over the counter as possible. This is now rapidly developing with online sales models with basically no people involved. Delivering goods, lasting quality, and solid sustainable products have become less important or completely absent.

Today, a marketing strategy (go-to-market) is more important than a good, solid product. If consumption can be created, it does not matter what consequences the production or the consumption of the product has on the environment, society, or the climate. This has resulted in a great deal of creativity in the marketing of the products with storytelling as a predominant trend. In many ways, over time, storytelling has become a set of banalities in trying to market similar products produced far away. With the global value chains, the customer has moved away from seeing and understanding production.

Companies are willing to sell almost anything, and delivering high-quality products is done by the few. The principle of OEM (original equipment manufacturer) has arisen from this linear way of businesses thinking. OEM is a method where products are being produced at mega-factories a few places in the world, a long way from where they are consumed. The companies in the end of this value chain are only providing marketing and sales to feed the *take-make-use-waste* business model. Typically, these products are manufactured in Asia. Whereas, it has been the marketing and packing of the products that has created the "uniqueness" (USP).

In recent decades, storytelling, as we know it from fiction, has moved deep into marketing, and when customers buy a product, a new car, clothing, watch, or whatever, they buy a philosophy of life or a whole new identity. Anyhow, that is what the marketing companies are trying to convince us of. This has led to many branding campaigns moving away from the product and almost becoming a story of a better and more real life. As sustainability has grown high on the consumer's agenda, the branding campaigns have also included this element—often hidden, but they try to leave an impression that the products are sustainable without changing neither the business models nor the products. In a Green and Circular Economy storytelling most be upstream about the supply chain and production of the product instead of downstream the value chain about the consumer and their lifestyle. Now, it is more important to understand how products are manufactured, where and from what, than to promote consumption with a "lifestyle" framing.

In the future, and more than ever before, marketing and sales activities will be much more dependent on the actual product, production methods, and material origin and recyclability. As well as dependent on the documentation of the product's ESG footprints in the full lifecycle.

The products produced as OEMs, the marketing, and greenwashing going on here can only happen because customers have not been critical enough in understanding the products' global value chains and in questioning quality, production methods, and environmental impacts. This is changing these years with young, talented users and an increasing number of critical consumers at all ages that question where and how the products are produced. There is a growing criticism of company greenwashing, and this criticism will only increase in the future. Local commerce regulation and EU legislation on green claims will be the end of greenwashing of products and corporations.

In the long run, companies—especially among global corporations—will not be able to leave an impression or give a narrative of being sustainable or complying to the ESG sustainability standards without accounting for the impacts. By legislation, companies must document the changes that make them sustainable, and sustainability will become the new business purpose. Therefore, it is essential to properly tackle the transition of the business and its products in a serious manner and to dive deep when doing so. Just as certification and documentation are becoming more important than ever before, new initiatives and standards are coming from the EU together with a product due diligence for imported products.

Human resource management remains a support activity and will mainly be challenged by the fact that companies need to cooperate closely with the full value chain/supply chain in new ways. Rather than where companies are very specializing within one step or silo of the value chain. On the other hand, more specialized technical and sustainability competences will be needed in the companies. When selling the product, the salesperson must have a thorough knowledge of the product life cycle, and she must negotiate the terms for take-back or recycling of the product. Procurement will become a much more complex activity with agreements covering sale as well as take-back and recycling of the product. Including relations with the new material banks for purchasing of recycled materials, etc. Therefore, sales and procurement departments and their activities will fuse together in the circular organization. The transition to a Green and Circular Economy is the battlefield of the nerds and a new technical journey. This set new requirements to the future employees that HR must understand and monitor.

Human resources will be challenged to recruit the best employees with new competences within sustainability and responsible business in the future, especially on management levels, and here a strong sustainability profile of the company becomes essential. HR will also face new challenges in implementing the circular business models and the circular organization in the existing companies and with the existing employees. The future purpose of the companies will be to have an anchor in society with an extended responsibility for society, people, and planet. Thus, taking care of the long-term interests of its owners.

The most important is that the great transition starts with a completely different composition and different competencies of management to change the organization, the products, and the business models. Currently, there is a lot to talk about sustainability and the SDGs in management corridors, but there is still a long way to go before the transition becomes strategic and business oriented. And a long way before management takes the full responsibility in the extended value chain, which also causes changes in the organization of companies.

In this new version of Porter's value chain, the circular organization illustrates how companies must operate in the Green and Circular Economy. First it is important to educate and recruit new competences that can drive the change and fill in the new job profiles in this organization and that will become essential for the future corporations. Read more about this in chapter 4 of this book.

For many years, marketing has been the king, convincing customers of a need for almost anything that then could be manufactured. In the future, the importance of the activities in the company's value chain will be turned upside down. The development of products into the Circular Economy will be technologically driven rather than customer driven, as we know it today. Overall, the circular requirements from the EU and the sustainable product regulation with new eco-design criteria and a product environmental footprint (PEF) will demand transparency and traceability in the full value chain of the products. It will result in product prices to increase because of the new extended producer responsibility forming a Circular Economy, followed by many new product requirements. The new business models will include take-back systems of all products—directly or indirectly. Either as product or as disassembled materials. Implementing these new recycling systems require investments to establish and run but, on the other hand, result in circulation of materials that give access to secondary materials becoming cheaper in the future than virgin materials and ensure the future supply of materials. All in all, companies need to reorganize to meet all these requirements.

Companies and a Green and Circular Economy are driven by access to sustainable materials and several regional economies closer to local communities rather than one global economy. To some extent, the Green and Circular Economy makes up with globalization—even though it is not anti-globalization. We need to take care of our resources and our planet, and then the new economy will be based on processes closer to us. So, proximity to products, materials, the recycling loops, and our nature is an important driver in the Green and Circular Economy.

References

Haar, G. (2024a). Chapter 6: EU regulation to a green economy. In G. Haar (Ed.), *The Great transition to a green and circular economy*. Springer.

Haar, G. (2024b). *Chapter 9: Transition to a circular economy*. Springer.

Haar, G. (2024c). *The Great transition to a green and circular economy*. Springer.

Porter, M. E. (1985). *Competitive advantage: Creating and sustaining superior performance*. New York: Free Press.

Part II
Sustainable Development Goals for Transition in a Nordic Perspective

Part II—*Presents the Sustainable Development Goals (SDGs)*—as a framework for business, organizations, and societies in a Nordic perspective. Here is a view from and on the Nordics as societies and the Nordic businesses including an introduction to the Nordics to understand some of the characteristics of the societies here.

This part helps leaders but also others to understand the SDGs and provide new ways of accessing businesses in a sustainable context, providing insights on the urgent need for change and how sustainability and new economic solutions will become the largest strategic challenge that companies will face now and in the future. The SDGs are organized in a Strategy House to assist leaders in all levels of society to prioritize the SDGs. This book reviews the SDGs and puts them into perspective of the Nordics for companies and others as a central wheel for driving the changes also on societal level.

This Part II is a review of the SDGs through a Nordic lens and this chapter is a short introduction to the Nordic countries as an introduction to the other chapters in Part II.

- Companies
- Municipalities and public institutions
- Society model

The chapters here are general guidelines to understand the SDGs and use them as a navigator for nations, societies, and businesses to be stakeholders and drivers in the transition to a fair and sustainable planet that is needed to save humanity. Lots of other models and frameworks are available as:

- ESG—Sustainability Standards
- UN Global Compact
- Science Based Targets
- Others

All these frameworks do not exclude each other, they overlap and many stand on the SDGs.

Chapter 7
Introduction to Sustainable Development Goals and the Nordics

The UN Sustainable Development Goals (SDG) have become a new focal point for environmental as well as social sustainability. Nations, organization, and companies are using it as a framework for communicating sustainability, visualized in Fig. 7.1 (UN, 2024). They have attracted considerable attention, especially in the public sector and among businesses, here in the Nordic region. Politicians have begun to adopt the SDGs, while generally it is still only the very dedicated people who investigate and are familiar with the content of the SDGs. Some people have heard about the SDGs but very few people know the 17 goals, the 169 targets, and that they are due by 2030. The SDGs are not yet creating the large, global movement that they were hoped for. Even though they contain and describe the challenges and solutions that people and the planet are in the need for on a larger scale.

As mentioned earlier, the Nordics are getting a lot of attention these years from all over the world, as an attraction to the Nordic lifestyle, design, and society model. The Nordic societies have proven resilient and innovative through the global crises since the evolvement of a strong global economy in the last three to four decades. Many companies and institutions in the Nordics are using the SDGs in their communication of sustainability, and the SDGs resonate in the Nordic way of thinking and doing business. Still very few uses the SDGs as a map for strategic development of their business models or in their organization which is necessary if they are to drive change. Many companies, also in the Nordics, are still challenged on incorporating them into their core business and into their value (supply) chains at a strategic level.

First it is important to read the fact sheets attached to the SDGs, as well as the targets and indicators that unfolds them to get the necessary perspective on the challenges they address. All the targets and detailed material can be found on the UN websites, where also small videos and facts are introduced.

© The Author(s), under exclusive license to Springer Nature Switzerland AG 2024
G. Haar, *Rethink Economics and Business Models for Sustainability*,
https://doi.org/10.1007/978-3-031-56653-0_7

Fig. 7.1 **Sustainable Development Goals (SDGs).** In understanding the Sustainable Development Goals (SDGs), it is important to investigate the targets and the educational material provided by the UN. This material presents the global issues in creating a fair and sustainable world in a good and descriptive way. See https://sdgs.un.org/goals/

Introducing the Nordic Countries

The Nordics includes the five countries—Finland, Norway including Svalbard (a group of Islands far north), Sweden, Iceland, and Denmark including Greenland and Faroe Islands. The Nordic countries hold a strong social and historical connection. Of the five countries, three of these make up the Scandinavian countries: Norway, Sweden, and Denmark that once was one kingdom. Dating back to the twelfth and thirteenth century, the Scandinavian countries were united in the Kalmar Union by Queen Margrethe I., who ruled on behalf of her son the King and married to the Norwegian King Haakon II. She understood that the Scandinavians needed to stand together and become allies with England (Britannia), to stand against the strong German-Roman Reich of the South. She formed the Kalmar Union that since created a strong historic and social connection and is the basis for unifying the countries in the Nordic Council today. It is important to mention that even if most of the Nordics are constitutional monarchies, the royals do not hold any significant political or economic power. They have a social power and are generally popular among the people and provide a kind of cohesion to the people. They are the formal heads of the military and the church, and the actual power lies with the parliaments of the Nordics. The royal's main task are to be good ambassadors for their country and the businesses.

A map of the Nordic countries and their flags are seen from Fig. 7.2, and in Table 7.1 are some facts on the Nordic countries. The flags of the Nordic countries show that they are all Christian countries that were christened approximately 1000 years ago. All the countries have Lutheran Evangelical People's Churches as the dominating religious communities. Most of the people of the Nordics have a very balanced and relaxed approach to their religion. Due to the high level of education all through the Nordics, a science-based and fact-based approach to life and the state is dominating.

The Nordics all share the cold and dark weather and the ability to adapt to a life situation where food can only be grown in a short period of the year and where sheltering from the tough weather is a strong part of the culture. Preserving food, livestock, and housing was also a strong part of the tradition and the cuisine here. In the last decades, the influence from the USA and Southern Europe has been strong on cultural development, but it is as if the Nordics are redefining their cultures back to the traditional roots these years. As seen from the map in Fig. 7.2, the arctic circle goes through the northern part of the Nordics, and above the circle people live in complete darkness for a period during the winter and at full sun light through the day and night during a period of the summer. This affects the mentality and the moods of the Nordic people and creates a certain cynicism, efficiency, and withdrawn personality that is so characteristic for the Nordic people. It requires a certain stamina to survive these conditions.

The free thinking and liberal personal freedom are a backbone of the Nordic societies that foster even and equal access to education, healthcare, and social welfare to support the individual freedom. From outsiders this is often considered as socialism even though the market and business environments here are based on a capitalistic and liberal market model, with a strong social welfare.

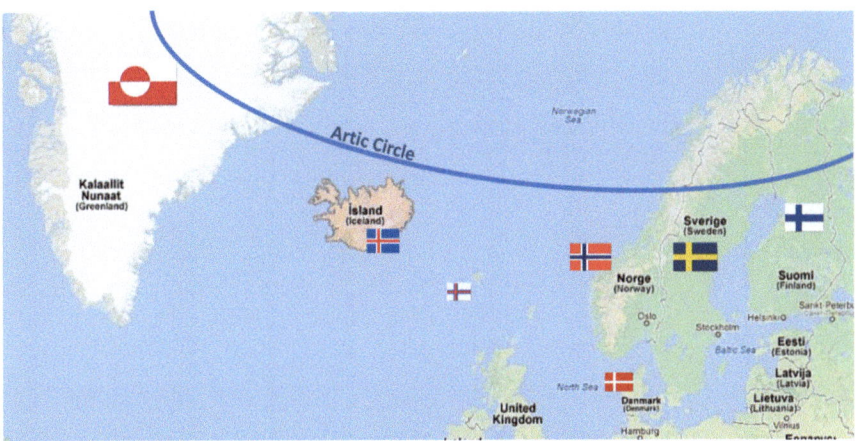

Fig. 7.2 Map of the Nordics. A map of the Nordic countries including Greenland and Faroe Islands. The arctic circle is indicated on the map as are the Nordic flags

Table 7.1 The Nordics. The GHG emission stated is scope 1+2 only

Country	Population (mio)	Area km2	Capital	Characteristics	GHG / citz (t CO2 - 2021)	Governance	Flag(s)
Finland	5.5	338,145	Helsinki	Forestry, lakes. Young democracy. Nuclear, hydro and wood power	6.8	Republic	
Norway	5.5	385,207	Oslo	Mountains, fjords, fishing and rich on oil. Hydro power.	7.6	Constitutional monarchy	
Svalbard	0.003	62,050					
Sweden (3. largest in EU)	10.5	450,295	Stockholm	Forests, rivers, industrial, and neutral until 2023. Nuclear and hydro power.	3.4	Constitutional monarchy	
Iceland	0.4	103,000	Reykjavik	Volcano island, fishing and innovative. Wave and geothermal power.	4.0	Republic	
Denmark	5.9	42,943	Copenhagen	Agriculture, fishery ending, biotech, many islands, self-supporting on oil and gas. Greenland rich on minerals. Faroe islands fishery	5.1	Constitutional monarchy	
Greenland (autonomy)	0.057	2,166,086	Nuuk		9.7		
Faroe Islands (autonomy)	0.053	1.396	Thorshavn		4.0		
Total (size of Peru or larger than India when including Greenland)	28	1,383,036 excl. Greenland	-	Cold, dark and far north with oceans, water and primary production	-	Historically and socially tied together. Politically in the Nordic Council	

A special management style and business culture have evolved in the Nordics that receive attention from the outside, as inspiration to make companies more agile and to implement the changes needed. As described in Part I, this is characterized by the lack of power to authorities in the daily business life, an informal tone, flat hierarchy, and the ability and obligation to speak out against authorities.

Maybe the Nordic countries are where to look for societies and business environment that are able to implement the SDGs. This is also why this Part II of the book processes the SDGs in the backdrop of the Nordics hoping to inspire and give a way into cracking the nut of implementing the SDGs.

Chapter 8
Sustainable Development Goals in a Company Perspective

The SDGs are mainly written for nations and for the politicians and institutions of society to create the structural changes of society and legislation to meet the major global challenges, as inequality, hunger, education, sustainable ecosystems, and innovative development of cities, business framework, and infrastructure. Again, this is great educational material and video, as well as the targets, events, publications, and actions, are available free and online at https://sdgs.un.org/goals. This information is available in most languages.

SDGs Are also ESG

ESG is a well-known framework for companies and also the framework in the EU legislation on disclosuring on Sustainability. This does not out rule the SDGs and what they offer to companies. The SDGs have a built-in color code that fit with ESG and the way many companies and institutions look at sustainability. As stated here:

- The blue and green targets are the environmental challenges.
- The reddish targets are aimed at people and the human rights challenges.
- The orange/brown goals cover economic structure and infrastructure.

Looking into the SDGs, it may seem surprising that SDG#2 is light brown and not red. That is because hunger is a political, economic, and structural issue. There is enough food available on the planet. The distribution is what keeps people starving. The hunger problem must be solved by changing political and economic structures. Not by growing more food on a global level.

EU has defined sustainability as ESG in the legislation covering Environment, Social, and Governance (Haar, Chap. 6: EU regulation to a green economy, 2024a). Financial institutions and stock exchanges all over the world are requesting inputs on the ESG topics from companies. The USA is also introducing ESG reporting on a voluntary basis. The US Securities and Exchange Commission (SEC) only requires companies to report on information that may be material to investors, which includes ESG-related risks. The USA is split on the ESG disclosure requirements as 22 states that have adopted some form of ESG-related laws, 18 states that have adopted "anti-ESG" legislation, and 14 states that adopted "anti-ESG" legislation in the 2023 legislative session. Still, it seems as if the financial partners have an increased focus on ESG as it impacts corporate performance on the long-term all over the globe (2023).

SDG and ESG are confusing to many. In a simple way, the ESG is overlapping with SDGs as illustrated in Fig. 8.1. The EU Commission stated that the Sustainability disclosure regulation is built on the Sustainable Development Goals. This means that transferring between ESG and SDG should be easy.

Businesses are expected to use the ESG sustainability standards as this complies with various disclosure requirements. This should not hold companies from using the SDGs as they are very educational and good to communicate against. Realizing that the SDGs are not written for businesses, business activities have a great influence on the progress of the goals, and the framework is unfolding the sustainability and impacts issues to a much larger extent than ESG requirements. Some companies already communicate actively using the SDG framework especially in the

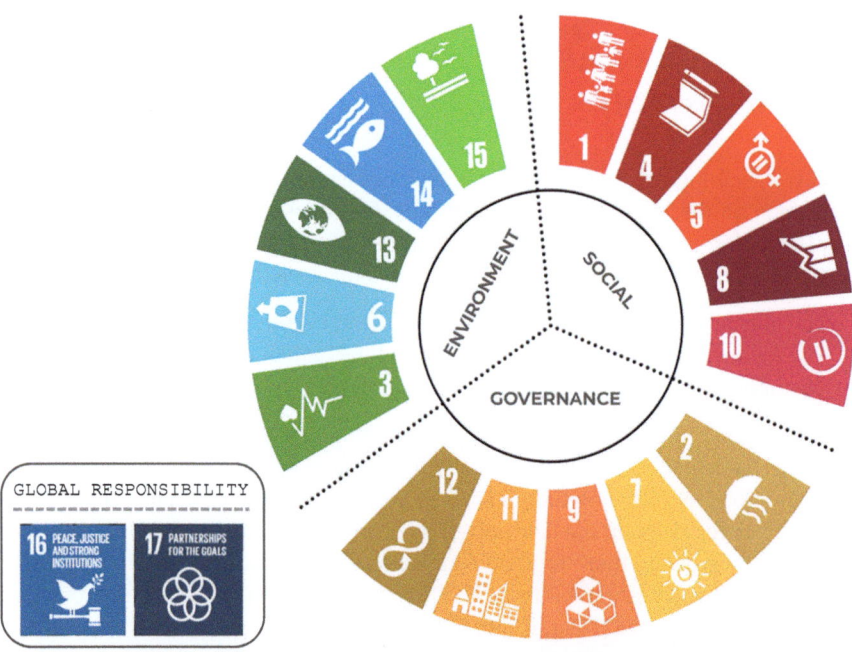

Fig. 8.1 SDGs are ESG. Many companies and organizations in the Nordics use the SDGs for communicating sustainability. Now legislation is imposed on companies referring to ESG as the framework for sustainability. The ESG Sustainability Standard complies good with the SDGs indicated by color codes

Nordic. Although many companies are struggling with how to implement and work strategically with the SDGs.

SDGs as a Strategic Tool

Here, the SDGs are presented as a strategy house that can be used to identify the relevant SDGs for companies, organizations, and law makers. Then the SDGs are reviewed in the perspectives of companies, organisations and society in the chapters in this Part II. Part III of this book includes a guide on how to get started on the sustainability journey as a company. Here a Sustainability Roadmap is introduced, as a tool on how to manage the long strategic journey that most companies need to engage in. The selected SDGs may be used in the Sustainability Roadmap as the communication framework, as well as the ESG. The strategy house helps prioritizing the relevant and important SDGs for the specific company or organization to build a strategy and development of new business models, policies, actions, and monitoring the company and its value chain.

The SDGs support the great transition to a Green and Circular Economy that companies are increasingly to prepare for. It is important to respect the urge and significance of the SDGs and not to dilute their mission for a fair and sustainable world. Just using the SDGs as a communicative play of cards by mapping the existing work to SDGs is disrespectful to the challenges they address as well as the challenges that the world is facing these years. Then it is just a new way of greenwashing.

Some companies or organizations (private or public) believe they must embrace all 17 goals. Then they have not read and understood the targets and that this is a strategic platform rather than a communication platform. The SDGs present an agenda of change, called the Agenda 2030. Strategy and change management requires focus and depth why companies only can work with a few selected SDGs, and these are the SDGs closest to their business. Embracing all 17 SDGs is a misunderstanding of both the leadership task and the SDGs. Implementing a new strategy does not contain 17 key elements or 17 Must-Win-Battles. The number of MWB in a new strategy is 3–4 at the maximum. The same goes for the SDGs, and it is important that companies select only a few SDGs and then work strategically with these. The recommendation is to select between two and five goals that are close to the business models and close to the impacts in the full value chain. A clean tech company providing water purification installations chose SDG#6—clean water. The textile industry chose SDG#8—responsible jobs and economic growth.

The Tool

In Fig. 8.2, the SDGs have been put into a strategy house, which is a method business leaders and management consultants often use to visualize company strategy. This illustrates that the SDGs contain visionary goals, tactic goals, and enablers.

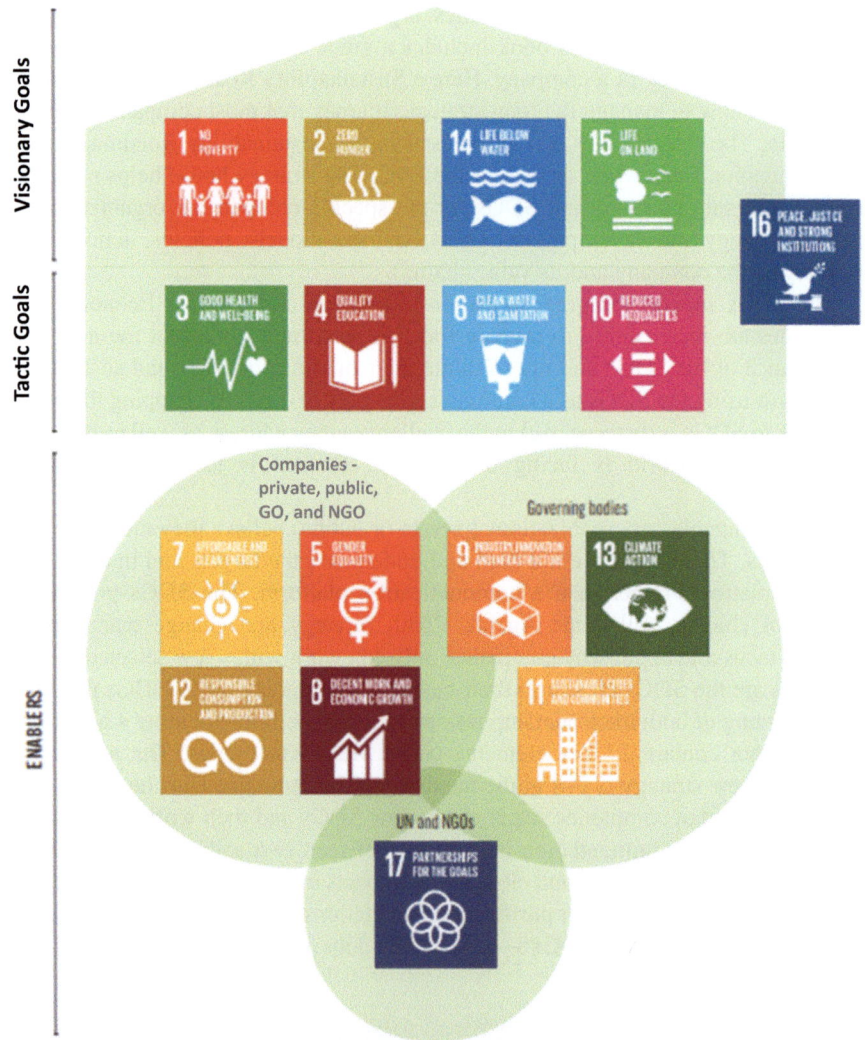

Fig. 8.2 Strategy house. The SDG strategy house to help politicians, municipalities, companies, private and public, government organizations (GO), and non-government organizations (NGO) to prioritize SDGs

Visionary SDGs

The Sustainable Development Goals are built on some overall goals, the four overall and visionary SDGs. They form the ground for a sustainable world for people and the planet. The SDGs is a universal call to action to end poverty, protect the planet, and ensure that by 2030 all people enjoy peace and prosperity. Regenerating natural ecosystems on land, in oceans and in fresh water, is necessary for nature to absorb greenhouse gases instead of emitting them, and to counter climate change. In contrast to the increasing cultivation of land and the logging of forest these years causing loss of biodiversity and GHG emissions. Just as the increasing temperatures and salinity of the oceans change the ecosystems, and coral reefs, plant life, and wildlife become extinct. The oceans and wild nature are no longer the buffer capacity for GHG that counteracts climate change, as they have been historically. Regenerating mangroves, coral reefs, and wetlands will make the ocean to absorb CO_2 again and importantly act in mitigating climate change.

These four goals are overall goals and the other SDGs support these overall goals to be met by 2030.

Tactic SDGs

The four tactic SDGs or sub-visions (SDGs# 3, 4, 6, and 10) are humanity goals that shall bring fair and wealthier lives to people, globally. These SDGs are about creating the basic conditions for a decent life and a basis for the visionary goals to come through. The tactic goals connect to the original objectives for establishing the UN back in 1948, all built into the universal human rights. However, the global populations were not able to reach human rights for all over the last 70 years, and the SDGs are a way of realizing that the state of the planet is essential for spreading humans' rights globally. The tactic goals are a lever to a visionary goal and a fair and sustainable planet.

Enablers

The two pillars in the strategy house are the enabling goals and are visualized in overlapping "balloons" or circles. Overlapping because these SDGs must be solved jointly between many stakeholders in society. Citizen, companies, states, municipalities, NGO's, etc. Meeting the SDG enabling goals will create the change with involvement of the businesses, the local governments and municipalities, and institutions by engaging people on local levels. It is in the interaction between local governments, regions, municipalities, and businesses that the change must happen because this is where people work, live and act. Businesses are key players in addressing the SDGs right where it matters. Businesses are made up of people and deliver to people. Local and national governments are elected by people and create the framework and infrastructure for people. So, the interactions here are significantly important. The future presents a new type of business humanism, where companies take greater responsibility for their impacts on people and planet. Integrating a purpose and a greater responsibility in businesses in interaction with society will change the way we all interact, whether as employees, managers, and customers, or otherwise in civil society.

The strategy house in Fig. 8.2 helps companies, organisations, and local governing bodies to prioritize the SDGs that are linked to the core of business and conduct. This chapter focus on assisting companies in their strategic work with the SDGs, and the following chapters address other governing bodies and society in general.

All companies in the Western world must first relate to six important SDGs. Within these six SDGs virtually, all companies can make their significant difference. Not only to the business but also to the planet and the people in their global value chains.

The six important Sustainable Development Goals (SDGs) for businesses are:

    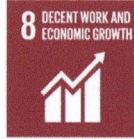 

Responsible Consumption and Production

These six essential business SDGs are reviewed here, and background information available online on https://sdgs.un.org/goals/ by the UN in October 2023 has been added to this review.

This is the most important goal for businesses in the old, industrialized countries, especially in Europe, and partly in Northern America, called shift countries in the Circularity Gap Report (Initiative, Circularity Gap Reporting, 2022). The overconsumption generated here creates many of the challenges and inequality that exist globally. This relates to the challenges addressed by the SDGs in the developing countries world, called build countries in the Circularity Gap Report (Initiative, Circularity Gap Reporting, 2022). The people and societies in the old, industrialized countries have since World War 2 created a way of consumption that has escalated even further in the recent decades. Today, there is a consumption of all types of goods, as clothing, electronics, long-lasting consumer goods, and building materials that is quite extraordinary and create large challenges due to the linear business models. Today, products are manufactured for disposal and with the only purpose of selling as many units as cheap as possible, and with no respect of the drag on raw materials and manufacturing methods. This is why SDG#12 is the most important SDG for companies providing for the consuming part of the population and addresses companies from the old, industrialized world.

Engaging in SDG#12 is where companies are going to make the most significant changes and that will affect their business models the most. Limiting consumption and transforming to a Circular Economy where all products and materials are reused and recycled, and the value and materials built into products and society are maintained infinitely. Only then we can maintain the prosperity and put a responsible footprint on the planet. The Nordic countries have highest consumption and highest production of waste, globally. The high ambitions of the EU and the Nordic Council, and the goals and strategies on mobilizing the industry for a clean and Circular Economy will hopefully make this change.

Even more embarrassing is that in our eagerness for consumer goods and new stuff, we have given up on the quality of what we purchase and consume. For a great many of the consumer goods, the quality today is significantly poorer than just 10 or 20 years ago. And much poorer than 70 years ago. This means that we destroy the ability to keep resources in future loops. Good materials are polluted with a lot of harmful chemistry, and materials are mixed in ways that make it difficult to maintain and recycle them. The examples here are obvious: building materials, textiles, furniture, food, electronics, etc. All these products fall short on quality, and consumers have ignored that the manufacturers flew off at a tangent in our eagerness for cheap stuff.

The EU Green Deal and the Action Plan for Circular Economy will change the market conditions for companies in the transition to a clean and Circular Economy. The crucial elements here are the Sustainable Product Regulation (SPR) and the Extended Producer Responsibility (EPR) that will implement sustainability in the full value (supply) chain, as well as reuse and recycling requirements, new eco-design criteria, and chemical legislation. This is followed by requirements on labeling of all products with a product environmental footprint—PEF carried by a Digital Product Passport (DPP). All, which will be central in creating responsible consumption and production again. In particular, the new EU product labeling (PEF/DPP) and the new regulation on ESG-disclosure can form the basis for taxing the negative impacts from companies on the environment and the exploitation of externalities. A few EU countries have started taxing different product categories, and this is likely to spread rapidly. Read more on legislative framework in another book of the same author (Haar, Chap. 6: EU regulation to a green economy, 2024a).

Discussions on who is to blame for this overconsumption are irrelevant, even if these discussions are running among politicians and in the UN. Is it the consumers who have been blindfolded or have the producers manufactured just what the consumer demanded, against knowing better? Most importantly, is it that the framework for manufacturers and consumers to switch to a clean and Circular Economy is now created in the EU with China following rapidly and a few corporates in the USA. A Circular Economy requires completely different designs, quality of products, and completely different possibilities and requirements for maintenance and repair. This result in new business models, new material loops based on reusability and recyclability (Haar, The Great Transition to a Green and Circular Economy, 2024b).

Almost all existing business models in the old, industrialized countries are based on growth and huge consumption of goods that is replaced when or before they are *worn-out*. The purpose of this linear business model is to have consumers change to new products rather than to repair and maintain. Thus, the disposing of large values is happening without any benefit. This overconsumption is based on global value chains where goods can be sourced very cheaply outside the old, industrialized countries. Much of the wealth created especially in Europe, Japan, and in North America in recent decades, has been created by cheap production outside these countries. At raw material prices and wages far below the wages for those who consume the goods. It is a good idea to use the global value chains to create economic development on a larger scale—both here and in developing countries. These linear business models and the manufacturing methods are so irresponsible, both economically, environmentally, and socially, and keep people in poverty, and they do not belong in a fair, sustainable, and transparent world. If we are to respect the basic human rights, such as access to clean water, food, and education jointly agreed to by the countries, in their membership of the UN. Therefore, all companies should start with SDG#12 and strive to meet the Circular Economy and build a responsible value (supply) chain. This to ensure resources to be restored and recycled and promotes responsible consumption, especially when dependent on long, global value chains. Then a shift to responsible economic growth (SDG#8) is necessary to create the changes needed. More on this later.

Affordable and Clean Energy

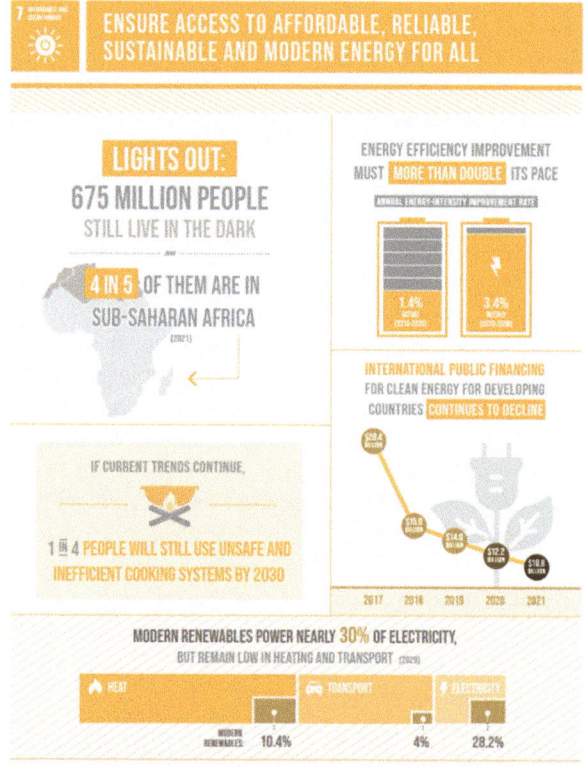

THE SUSTAINABLE DEVELOPMENT GOALS REPORT 2023: SPECIAL EDITION- UNSTATS.UN.ORG/SDGS/REPORT/2023/

Other major challenges for companies are climate change, climate adaptation, and affordable energy supply. It is important how we all contribute to reducing the GHG emissions that cause the climate change. The good news is that investments in energy savings and renewable energy are profitable for companies. So, all companies should ensure renewable energy supplies as soon as possible, and use no more energy than necessary. There is an estimated potential for energy savings in the industry sector of 20–40%, depending on the industry. The potential is the same for energy savings in many office buildings, especially in the newer, glass facade office buildings that have risen in great numbers in the last decades, because they need severe cooling. These potentials are often overlooked, and they are pure cost-cutting projects leading to saving GHG emissions. Read more about energy optimization in companies in another book of the same author (Haar, The Great Transition to a Green and Circular Economy, 2024b). All this is also very relevant for municipalities and states that often own larger number of buildings, and hold the same potential as the private sector. In Part III of this book, a method is presented to drive energy optimization including installation of Renewable Enery useful for all types of public and private sectors and their portfolio of buildings.

In general, the countries in Europe and especially in the Nordics declined their fossil energy consumption over the last decades. Mainly due to transformation to gas supply being more energy efficient than coal and oil, due to installation of renewable energy, and to some extent due to energy efficiency in buildings and in the industries. The electrification of the industries that is necessary to switch to renewable energy also offers energy efficiency, as the electric engines are almost double as efficient as the combustion engine. This efficiency needs to be harvested as soon as possible. In general, there is still a potential to cut down energy consumption in the full life cycle of many products and companies by electrification and by recycling of materials, as recycling is much less energy intensive than extraction of virgin raw materials and processing of these.

The buildings, private, public, and industrial still hold a potential for energy optimization. Here it is very important to look at the full life cycle of a building before retrofitting as up to 70% of the climate impact from a building relates to the production of building materials and the construction and demolishing of the building. Not as many believe the energy consumption in the operational phase that only accounts for up to 30%, today. In the future with intelligent management systems and renewable technologies, buildings become energy producers, and then the climate impact will only come from how we design, manufacture, and construct buildings, and the old, existing building stock, some dating centuries back, does only add a smaller climate impact from operating the building. Retrofitting to minimize the energy consumption for operation can have the opposite impact in the full life cycle. Therefore, managing climate impact in the building industry or as a building owner requires the understanding of the Circular Economy and the life cycle of building impacts.

In Europe and in the Nordics, energy optimization programs has solely focused on minimizing energy in operating the building without accounting for the negative impacts caused from exchanging building materials and the impact from the upstream supply chain, which is often outside the region. Read more about Circular Economy and energy optimization in another book of the same author (Haar, The Great Transition to a Green and Circular Economy, 2024b). With EU's strategy and action plan on CE and the new strategies towards the building sector (FIT for 55, New Bauhaus, Renovation Wave, EDD) and national regulation to include full life cycle assessments (LCA), this is changing.

The EU has agreed on an ambitious energy efficiency target of reducing final energy consumption by at least 11.7% compared to projections of the expected energy use for 2030. (See https://energy.ec.europa.eu/topics/energy-efficiency/energy-efficiency-targets-directive-and-rules/energy-efficiency-targets_en.) The USA is very low on energy efficiency rankings compared to the EU (see https://www.energy.gov/scep/national-action-plan-energy-efficiency). The USA has released a "National Action Plan for Energy Efficiency that was a private-public initiative to create a sustainable, aggressive national commitment to energy efficiency through the collaborative efforts of gas and electric utilities, utility regulators, and other partner organizations. Such a commitment can take advantage of large opportunities in U.S. homes, buildings, and schools to reduce energy use, save billions on customer energy bills, and reduce the need for new power supplies." This is realizing the need for change, and it is to be seen what will happen here, but no doubt that the potential is large in the USA to create energy efficiency in the whole value chain of energy. China is the single largest consumer of energy, and a lot of the energy production is based on coal, but China's progress in implementing mandatory energy efficiency policies over the past decade has made it the world's energy efficiency heavyweight. China is delivering a lot of renewable technologies, as solar panels (EV), wind turbines, electric vehicles, and batteries. China is also world leader in transition to renewable energy, and with their centralized and efficient approach to targets, they will beat both EU and the USA on this. They see the transition to renewable technologies as their global marketplace.

The Nordics also have some global providers of energy efficient technologies, and due to the rough weather and an old building tradition, energy efficiency has been high on the agenda in most of the Nordic countries. Velux, Grundfos, Danfoss, Storaenso, Lindab, etc. In Norway Statskraft is a state-owned company providing renewable energy and driving the energy transition, as Norway export most of their oil extracted from the North Sea. Norway is investing money from the State Oil Fund to transform the building industry to a clean and Circular Economy and are active in the EU working tracks, even if they are not a member of the EU. Iceland, Sweden, and Norway have traditions within heat pumps and geothermal technologies.

Energy optimization and RE are now so profitable that it is irresponsible management, if not already implemented or at least the potential has been identified. Sometimes the profitability of investments in energy savings is higher than from investments in the core business. Yet there are too few companies looking in this direction or even making the effort to uncover the potential.

Much of the transition in the Nordics comes from state investments in larger RE plants, but companies can not only achieve profitability by investing in RE supply themselves but also a stable energy supply. This is the way to make production climate neutral in scope 1 + 2. This can happen entirely locally, but also in collaboration with nearby companies or municipalities. If jointly investing in solar panels, wind turbines, or other RE supplies. There are examples of industrial areas being made carbon neutral by wind turbine guilds. Then the company's climate actions become very concrete and visible for employees, customers, and the neighborhood.

Good Health and Well-Being

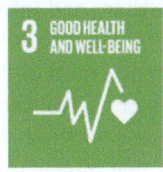

THE SUSTAINABLE DEVELOPMENT GOALS REPORT 2023: SPECIAL EDITION- UNSTATS.UN.ORG/SDGS/REPORT/2023/

The target here is to reduce the global mortality rates for all ages, which are distributed very distortedly globally. Mortality rates are a clear indicator of social inequality. The large targets in this SDG are birth mortality, child mortality, and the struggle against infectious diseases. This is what creates epidemics if not handled.

It creates very unstable communities when population groups become sick or die, and it influences many aspects of a society. The COVID pandemic and the way it was handled did put lights on the healthcare and public systems, and became a barometer for the quality of the systems and the responsible, well-functioning governments and sectors in the different countries. This also made corporates understand the value of operating in well-functioning countries in a situation, and that contributing to society within healthcare and infrastructure pays back in situations like a pandemic.

Some of the diseases that influence mortality in a country are birth mortality, AIDS, tuberculosis, malaria, and other overlooked tropical diseases, infectious hepatitis, and waterborne diseases. Viruses as Ebola, MERS, and SARS (as corona) can be counted as new type of diseases that infects from wild animals to humans.

In many developing countries, death and illness from road accidents take a much heavier toll than major diseases. A fact that is often overlooked. Just as hunger and diseases caused by migration of refugees are also very overwhelming. Business models and companies that rely on global value (supply) chains must pay extra attention to these new uncertainties that are both locally destabilizing, and also creating instable and insecure supply of goods and spare parts.

SDG#3 is both a tactic goal and an important goal for companies due to Target 3.9 that businesses should focus the most on.

> SDG#3.9 "By 2030, substantially reduce the number of deaths and illnesses from hazardous chemicals and air, water and soil pollution and contamination."

With the extended producer responsibility (EPR) that is now introduced in the EU and is well-known in the USA, the responsibility of hazardous chemicals in products and the surroundings now belong to the manufacturer. The risks from chemicals in products and production are increasing focusing these years in most industries. Medical studies are documenting the long-term consequences of the chemical impacts on human health to a larger extent than earlier. Therefore, companies should be more aware of the chemical content of their products than earlier. Including not only proven or listed banned chemicals but also chemicals with potential lighter health risks that may cause cocktail effects when mixed with other chemicals and ingredients. Chemicals will become a large topic in the future due to the risks to human health and a broader understanding of this.

In the Nordics environmental regulation and wastewater regulation has been strong for many years. Still scandals from early days appear today, and more will come.

SDG#3 also sets targets and indicators for diseases that humans inflict by overeating and smoking. In this way, the SDGs also include diseases that are widespread in the developed world creating inequality here. Companies can make an active contribution by the goods they put on the market and how they market them. The recommendations are active enlightenment and education to prevent lifestyle-caused diseases and to work actively to provide healthy food.

Another target in SDG#3 is the access to and the financing of healthcare as particularly important to overcome inequality. Here companies that source from developing countries should demand from their supply chain vendors that they provide access for employees to healthcare.

The global corporates, in cooperation with the UN, can demand that governments spend sufficient funds on ensuring the health and a decent healthcare system of their citizens and the employees of the local companies. Companies can play an active role in developing the healthcare not only of their own employees but also of suppliers and customers by entering binding alliances along the value chain to achieve SDG#3.

Gender Equality

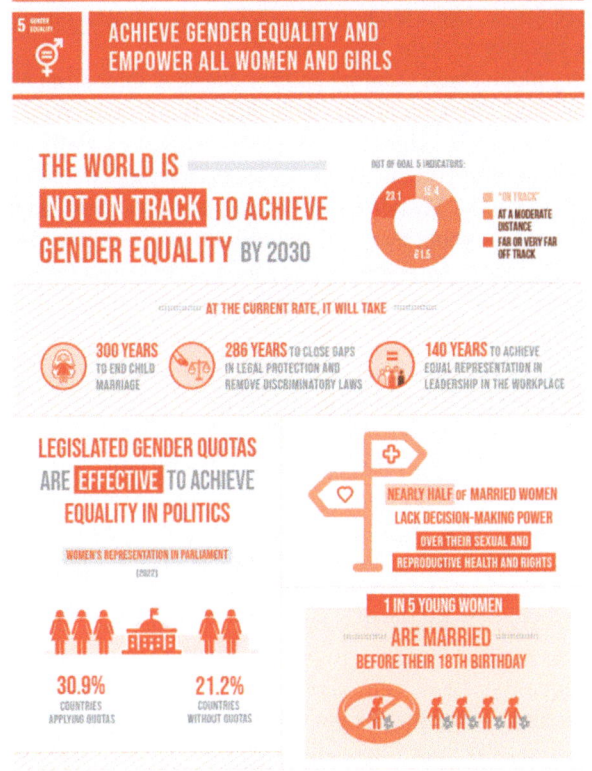

Gender inequality is a major global challenge for many reasons. Women rights are systematically oppressed in different ways in most countries around the world. Other aspects of this are becoming clear these years, and a great potential is lost when the talent and innovative power of women are not used to contribute to a better world due to historical structural oppression of women in many layers of companies and society.

Changing the world requires women to have a greater responsibility and opportunity to contribute to the development of the world. In a global perspective, many years of studies in development work have shown that if development aid is to work, it must be targeted at women. Women invest in the family, in small local businesses, and in agriculture. This creates more economic growth, education, and healthcare than when given to men. Gender equality parameters are now built into the development aid of most countries and programs. These learning should also come to the benefit of companies.

EU Commissioner for Development, Dane Poul Nielson (1994–1999), was one of the first who set gender equality as the focus for development, as he said: "If the world had been led by women, it would probably have looked different."

As responsible for the largest development funds in the world (EU), this is a very interesting observation. Unfortunately, it has brought limited change in development of gender equality over the last 20 years.

SDG#5 is very much about securing and protecting women from abuse, discrimination, coercion, and large economic inequality. Here, companies can influence their suppliers and customers through their value chains, especially outside Western countries, with binding requirements on gender equality and demands for transparency in how actors handle gender equality and issues on harassment. Companies can counteract this throughout their full value chain and with their own employees. Companies can also stop marketing women as a sexual object and especially the fashion industry but in general the display of women should have much more attention from companies and organizations. It is important that companies take responsibility and stop sexualizing young women in a way that substantiates this oppression. This is a wish from women all over the globe.

SDG#5.5: "Women must participate fully in management and decision-making" is a separate target and unfortunately very relevant in a Danish context. Although this applies broadly to management in large companies in most countries. In Denmark and the Nordic countries, women have achieved structural equality through legislation. In Denmark, there is still an inequality in pay for equal work, and in terms of gender equality in top management (executive and nonexecutive boards), Denmark is significantly behind, also compared to other Western countries and the Nordics. Repeated statistics published regularly prove this gender inequality. BCG and McKinsey have done several investigations within this area, and one of the results is shown in Fig. 8.3. (Boston Consulting Group, 2018).

As seen from Fig. 8.3, Finland and Norway are among the stars and Iceland and Sweden would be in the same area if included here. Denmark is in the field where

Gender Equality

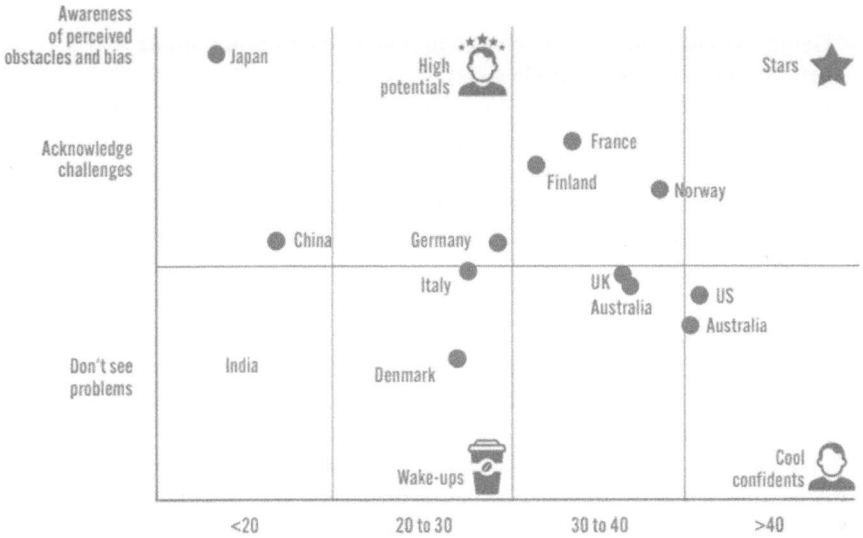

Fig. 8.3 Gender equality in management. Analysis of the experience challenges and bias by inequality. Minority as a share of total in % (2018)

only between 20 and 30% of management positions are held by women, and what is most remarkable for Denmark this is not perceived as a problem. Looking at the gender equality statistics for top management positions in Denmark, the number is far below the 20%. In C25 (25 largest public listed companies), there are two female CEOs equal to 8%. In 2020, 69 women were at the head of the 1000 largest companies equal to 7%. The situation on board level is even worse: in the 1600 largest companies, not a single woman chairs the board (Mogensen, 2020). In 2013 legislation was introduced requiring the largest companies to set targets for gender equality in management. Therefore, Denmark appears with a cup of coffee on the chart. It's time to wake up and do something about it!

There are plenty of women who want leadership and who have the skills, and there is an emerging focus on the fact that men have created a glass ceiling that prevents women from being recruited for top management (Mogensen, 2020). On the World Economic Forum's Global Gender Gap Index, Denmark ranks 95th (2018) on women in leadership. So, the free and liberal thinking does not go for women in leadership here.

> **Gender equality in companies and in management is not about satisfying the women, but because:**
> - Great business potential is lost when management does not reflect and understand their key stakeholders such as customers and employees.
> - The world and businesses are required to move towards a greater purpose and inclusion of ESG/SDGs and with extended responsibility for planet and people. This needs female values in leadership.

The BCG study (Fig. 8.3) also shows that gender equality is crucial to engaging employees. Companies with an engagement score of medium or less have only 8% women in management at senior positions. Conversely, the companies with a high engagement score among employees experience no gender equality gap, and their earnings are 20% higher (Boston Consulting Group, 2018).

A great potential is lost when corporate management is male-dominated, while customers, employees, and other stakeholders are equally gender-represented. In many industries, the female consumer is in excess. Some say that 70–80% of all consumer decisions are made by women. Therefore, it is reckless when large, global corporations do not work seriously with gender equality and targets for management level to mirror their customers and employees.

When looking at the distribution of gender of graduates from the universities and the grades that women achieve here, it appears that in most countries, women are better educated than men and have been for many years. Many companies recruit gender equally of young employees and newly educated. This is not reflected later at the management levels, not in the private sector, in public administration, nor in the educations and universities. Gender equality is changing rapidly globally, in the Nordics, but Denmark is starving for female leaders.

Many companies confuse gender equality and diversity, and this confusion does not benefit the implementation of any of these two topics. Both topics are important to solve in the creation of a stronger governance of corporations and to mimic the stakeholders of a company. There is a reason why gender equality has its own development goal, because the challenges and potentials of gender equality are at the same level at the other SDGs. Diversity to get representation of minorities in companies and in decisive bodies is also important, but women (or men) are not minorities. The two genders are almost evenly distributed in the populations. If gender equality is solved as if it was a minority issue, equality will not be achieved.

In the great transition to a Circular and Green Economy driven by ESG/SDGs and the change of impacts from companies on the planet and the people, these female values are needed. There is a fundamental and biological difference between men and women. Both as customers, employees, and business leaders. These differences strengthen the leadership of a company. Again, numerous studies show that the performance of a company is linked to its gender equality. That companies that have equality on management positions are more profitable and especially achieving better returns on their innovation efforts.

First, it is important to understand that many men possess the feminine values. Unfortunately, there seems to be a bias towards very masculine values in the male business leaders. Here a change is needed.

The feminine values in a work-related context consist of a greater focus on subject matter, creating change in a larger perspective; greater care for employees, culture, and customers; more focus on long-term solutions; and less focus on own, short-term financial gains. It is a major challenge that half of the global asset values are owned by very few men. The exorbitant salaries and bonus programs are given to executives given by men.

Gender equality is not solved by launching corporate female talent programs. If a large company lacks female leadership talent, it is the recruitment of employees and the lack of focus from executives that needs inspection. Not the talent pool of women available. There is plenty of talent among female employees, just look for it in the right perspective. They can and they will contest management positions. This is supported by BCG's latest study. Talent programs risk streamlining people and resembling existing leaders and mimic the existing management team, as well as slowing down the recruitment process.

So firmly put:

"We do not have gender equality until we see poor female executives in top positions."
Former CEO of the National Danish Bank, Lars Rohde. 2020

When companies take the SDGs, the megatrends, and the greater purpose of companies seriously, they must also take SDG#5 seriously. Make gender equality a strategic goal—preferably with gender quotas in management. It is not shameful to be chosen because of one's gender, as it is a quality to be a woman or a man. No doubt, that women and men possess academic and human qualities at the equal levels.

Decent Work and Economic Growth

THE SUSTAINABLE DEVELOPMENT GOALS REPORT 2023: SPECIAL EDITION- UNSTATS.UN.ORG/SDGS/REPORT/2023/

SDG#8 strongly follows SDG#12 because it also addresses the challenges of the global value chains.

In the Nordics, comprehensive legislation ensure labor rights, decent work, and decent payment for work. In a Nordic context, SDG#8 is about the inequality that exists in the global supply chains. When a company is transforming their business models and value chains to a Green and Circular Economy, they should look at the social aspects of their supply chains. They are responsible for creating decent work and decent economic growth all along the supply chain, especially if it contains production in developing countries; companies should to a much greater extent impose binding requirements for decent working conditions for own workers as well as the employees of their suppliers, and in the full supply chain. International agreements exist and have existed for a decade or more on social and labor rights (ILO, HR, etc.), but still much must be done. In the textile industry, current supply chains perpetuate inequality and social poverty to produce textiles at a quality and price that feeds the linear business models that creates global environmental challenges. Not only are the negative social impacts from the textile industry to be criticized, and the fast fashion industry must maintain inequality to be able to produce textiles at low quality and low prices to feed the linear business model. The textile industry is by many considered to be the second most polluting and climate-damaging industry after the oil and gas industry, especially causing harm to workers in the manufacturing end of the supply chain.

The textile industry and many other industries are challenged on labor rights and responsible economic growth in their value chains. The extended responsibilities that are put on companies in the EU, the extended producer responsibility (EPR) that are put on product for the EU market, and the due diligence requirements from the disclosure regulations will close the possibilities to source irresponsibly.

Another concern that companies must take seriously is the increasing inequality in the old, industrialized countries as, for example, in the USA and Britain. Both countries have high GDPs, but a very polarized spread of wealth. Both countries struggle with their democratic systems and with an impoverished working class that does not have access to decent healthcare, social security, or education. This has given rise to the concept of working poor, referring to families with two jobs that are not able to support a family. All along, the highly paid and the capital owners are getting richer and richer in the same countries. Thus, a polarization and economic instability is created that will influence the global economy and the financials of the companies in the long run. Decent jobs will become an increasing issue for companies that operate in the old, industrialized countries with increasing inequality. This is also why some leaders are looking towards the Nordic labor market and society model, as they now can sense the long-term consequences of the last decade's social responsibilities.

The wide economic disparities in many societies and globally are becoming so pronounced that 181 American CEOs declared that they want to benefit all business stakeholders in the future, including society, and that there is a need for a more equitable distribution of wealth in the USA (Roundtable, 2019). Read more on this in Part I of this book.

Peace, Justice, and Strong Institutions

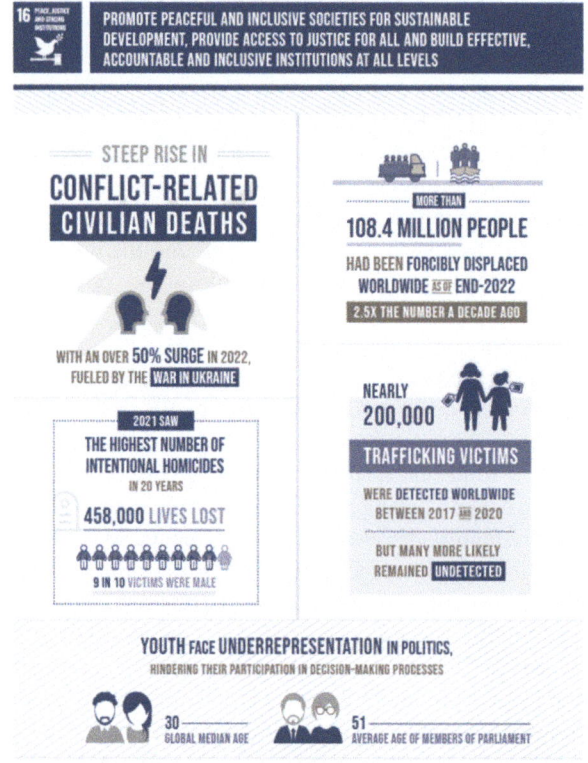

THE SUSTAINABLE DEVELOPMENT GOALS REPORT 2023; SPECIAL EDITION- UNSTATS.UN.ORG/SDGS/REPORT/2023/

Once the company has investigated SDG#12, 7, 3, 5, and 8, and identified how they can drive change, the next shadowing issue is corruption. The green transition and gender equality create great potential for businesses and for global economic growth—especially with an increasing potential in the developing countries. The developing countries have the potential to bypass some of the economic and environmental blind spots that the old, industrialized countries have experienced through industrialization the last century. However, there is an all-too-difficult challenge that creates and seems to continue to create great inequality, strife, and human misfortune. It is corruption. Anti-corruption is contained in SDG#16.

One may wonder that anti-corruption does not have its own SDG and is only mentioned as a target in SDG#16.5. SDG#16 clearly describes the need for judicial systems, impartial control, anti-corruption, the human rights, and institutions to promote and support peace and peaceful coexistence. Many companies have a code-of-conducts on this topic, and whistleblowers are set up to counteract corruption in companies. Nevertheless, it seems to be everywhere. Decency and trust are maybe the most scarce resources among humans.

It is surprising that a few of the Nordic banks have large challenges in keeping themselves from laundering the cash flows originating from corruption. Despite that the Nordics have strong traditions for transparent jurisdictions and a separation of the power of the jurisdiction system and the political power, and are known for trust, and little corruption. The Nordic countries are high on the list of least corrupt countries, published by the independent organization Transparency International. Moreover, the executives of these banks seem to have been aware of the money laundering for a long time. Few of the large banks in the Nordics can now reject being involved in money laundering activities. In relation to SDG#16, companies can send a strong message and introduce strong governance standards for anti-corruption and tax avoidance and not to interact and do business with suppliers and financial partners that do not take this seriously. The EU legislation imposed on the financial partners have become extremely strict in the latest years after numerous scandals in the sector have been revealed, and the large audit firms have changed their attitude in their advisory service within tax optimization after they have been count deeply involved in some of the fraud and laundry activities.

Partnership for the Goals

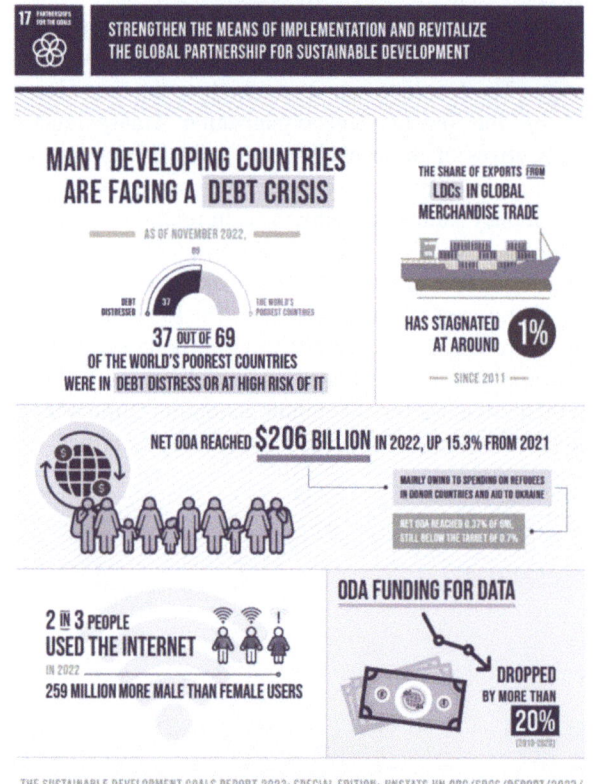

A lot of companies commit to SDG#17, and this is serious business to do so. Many companies put anything as a partnership and forget to read the target. SDG 17 is reviewed here, as it is certainly relevant for companies, but perhaps in a different understanding that companies use it today.

SDG#17 has gradually become a free ride for companies because everyone wants to cooperate about sustainability. This goal is not just about cooperation. It takes serious tasks to reach the SDGs and not something that is done with a few customers, vendors, or employee's workshops. Investigating all 16 targets under SDG#17 reveals that it is about strategic cooperation that strengthens developing countries.

Strengthening the tax collection and the tax revenues is one of the targets. If a company commits to SDG#17, it requires an extraordinary and far-reaching effort across borders and continents, where the company really reaches out and invests time and money. For example, as seen in the millionaire declaration, mentioned later in this book, and calls for property taxation and asset taxation.

The SDGs are only reached when the world has become fair and sustainable. Efforts are needed on many levels from citizens, to companies, NGOs, public organizations, and governments. Global efforts across and along corporate global value chains are crucial to success. The SDGs and in particular SDG#17 must be seen in this greater context and not as isolated actions for the individual citizen, company, or nation. SDG#17 is valid for companies when extraordinary efforts are made to change conditions along the value chains and when participating in changing legislation and framework conditions in the most challenged countries. The most significant effort with the greatest impacts is when changing the way people in the developed world consume and live. It is crucial how the products and services are produced because the global value chains are so important to people in all steps of the chain. Just as technological innovation flows across borders and can enhance our impact on the planet wherever we live, so can the way we manufacture and consume our products. Therefore, SDG#17 is obviously important for companies because we are all connected through the global value chains. Businesses should only be addressing this SDG if they are focusing on solving the global challenges in collaboration with all levels of stakeholders. The partnerships and targets addressed in SDG#17 are of a different nature than the partnerships many companies refer to.

Facilitating Tax Payments

Companies in the Western world can make a big difference and support responsible social development in countries with major social, environmental, and governance challenges by just paying local company taxes. For many years, tax optimization, leaning up to tax evasion, has been fully accepted by international corporations and rich private people, and to some extent supported by professional advisory companies. It is important to understand that a ridiculously small share of the values of global corporations and the richest few people can solve all the challenges addressed by the SDGs. The funds exit to solve the challenges in transforming to sustainability and promote economic development in the countries that needs to be built. It is the will that is lacking. Just as an example: in a few days, it was possible voluntary to collect enough money to renovate the Notre Dame after the fire in 2019 in Paris. These collected funds could have paid for the total cleanup of all plastic in the world oceans. The money was donated from wealthy private funds and private people.

Companies that adhere to the SDGs run a great risk if they also do tax optimization. Companies and shareholders that manage a responsible business and commit to the SDGs are also obliged by SDG#17.1. Companies must pay their taxes locally and contribute to create transparent financial transactions with customer and vendors, as well as anti-corruption. Then companies actively change the ability for countries to collect their tax payments and develop education, healthcare, and social welfare.

For example, the major dividend tax fraud that was revealed in Europe in the mid-2010 turned out to be organized through large reputable European banks and their investment funds. They actively offered an infrastructure where systematic fraud, tax evasion, and transferal of money to tax havens could take place. Such companies must not get away with using the SDGs in their marketing. This is lack of social responsibility. Decent tax payments have also become part of the Recommendations on Corporate Governance.

It is economic power of a few people and the consequent distribution of wealth that alone can change the living standards for all people on the planet. The first thing a company should do when working with the SDGs is to pay its taxes, and ensure that taxes are paid throughout the value chain. At the same time, governments and companies should agree on mutual requirements on transparency. Transparency as to how taxes are settled, applied, and spent. Here the UN, the World Bank, and the World Economic Forum should pay a more active role. Inefficient public systems and tax avoidance should be linked directly to the SDGs and ESGs. A requirement from companies should be that company taxes, and the income taxes of their employees are paid in exchange for responsible management of the tax revenue. Tax is a payment for companies to access infrastructure, justice system, and educated, healthy employees. It is therefore irresponsible not to pay for services received by society.

The rising inequality and lack of tax revenue to cover education and public healthcare is dawning on many, and 80 billionaires worldwide have signed a declaration to governments calling for permanent tax increases for themselves to address the world's post-COVID-19 challenges.

> "Today, we, the undersigned millionaires, ask our governments to raise taxes on people like us. Immediately. Substantially. Permanently."
> —Millionaires for Humanity—sign on letter.
> (www.millionairesforhumanity.org)

Therefore, it would be progressive if the UN, together with international corporations, were to put in a claim for governments and local authorities to comply with SDGs against corporations paying their local taxes. This could also be achieved through an UN-SDG fund funded by corporate taxes, which would ensure transparency in raising and managing the funds, especially towards solving the visionary and the tactic goals. Read more about a UN SDG Fund later in this Part II of the book.

All the eplanations and discussion on the SDGs in this chapter are in alignment with the requirements on corporations operating in the EU according to the Sustainable Finance Disclosure Regulation (SFDR) and Corporate Sustainable Reporting Directive (CSRD). The reporting regulation require more than highlighted here and it requires documentation on specific metrics and targets, as well as the implementation of an ESG-governancestructure in the company.

References

Boston Consulting Group. (2018). Retrieved from https://www.bcg.com/capabilities/diversity-inclusion/gender-equality/

Haar, G. (2024a). Chapter 6: EU regulation to a green economy. In G. Haar (Ed.), *The Great transition to a green and circular economy*. Springer.

Haar, G. (2024b). *The Great transition to a green and circular economy*. Springer.

Initiative, Circularity Gap Reporting. (2022). *Circularity gap report*. Circle Economy.

Mogensen, B. C. (2020). *De kan godt, de vil gerne (Danish)*. Gyldendal.

Roundtable, B. (2019). *Our commitment*. https://opportunity.businessroundtable.org/opportunity-commitment

Chapter 9
Sustainable Development Goals for Government Bodies and Legislators in a Nordic Perspective

Creating the conditions for citizens and businesses to participate and drive the changes required by the SDGs is the main purpose for legislators, government bodies, municipalities, public institutions, and state secretary offices. Their own organizations and public held companies must be managed in a company perspective as described in Chap. 8. Municipalities and state-owned companies or organizations must work actively with SDG#12, SDG#7, SDG#3, SDG#5, SDG#8, SDG#16, and SDG#17 when it comes to how operate their administration. These goals are even stronger imposed on the tax-funded companies than on the private sector, as they should lead the way and they operate solely for the purpose of securing decent lives and sustainable conditions.

Additionally, the public administration is implementing and monitoring legislation adopted by politicians as well as administrating and overseeing citizens and businesses to complying with legislation.

The purchase power of the public administration is often large. In Europe, this accounts for 50% of national consumption (spendings) and that put a special obligation on using this purchase power to drive the transition to fair and sustainable conduct by public procurement influencing the market situation. Large global organizations as the UN also hold a large purchasing power globally that they must administer according to the SDGs. The EU has announced and approved that member states must use public procurement to drive the transition to a Fair, Green, and Circular Economy. This is proceeding at a somewhat slow phase, also in the Nordics. Holland is an example of a country that is in the forefront of transforming the society and business environment to a Green and Circular Economy, and some countries around Europe are stepping up as Finland and France. But in general, requirements and tenders from municipalities and state administration are lacking behind in the demands for new, innovative circular solutions.

In the Nordics and in many other countries, the public administration is also a significant building portfolio owner since they often own the local kindergartens, schools, elderly care homes, sports facilities, and public administration spaces. The

© The Author(s), under exclusive license to Springer Nature Switzerland AG 2024
G. Haar, *Rethink Economics and Business Models for Sustainability*,
https://doi.org/10.1007/978-3-031-56653-0_9

building industry generates the most waste due to very little maintenance and renovation, and mostly demolishing and building new. The public administration as an asset owner financed by tax revenue carry an extra strong responsibility to drive the change and set future requirements to the whole industry, being the most significant provider and end user of materials and services in the building industry. Energy renovation and the switch to renewable energy is also proceeding at slow phase in the Nordics local administrations. Now, privately funded solar panel parks are emerging around Europe and in the Nordics, which contradicts very much with the scarcity of wild nature and biodiversity.

EU has a widespread framework of legislation covering the following strategies:

- Fit for 55 plans to reach climate neutrality in 2055, covering renewable energy, energy taxation, Energy Efficiency Declaration (EED), emission trading system, Carbon Border Adjustment Mechanism.
- Renovation wave in the building industry to become carbon neutral in 2055 in the full value chain including the existing buildings.
- Stepping up EU's 2030 climate ambition.
- Circular Economy Action Plan.
- Substantiating, sustainability claims.

All these change must be driven by the public sector and the state-owned companies showing the way for citizens and private businesses. Reality is that businesses are moving faster than the public sector in many places, also in the Nordics. The Nordic Council has set very ambitious targets on all the same topics as the EU, and now action is awaited.

On top of the obligations of the abovementioned SDGs that are compliable for companies, the public administration and the politicians hold a special responsibility of the three SDGs that set the framework for citizens and business to transform, locally. These are the SDG#9, SDG#11, and SDG#13, as illustrated in the right-side pillar in the strategy house in Fig. 8.3/9.1. These three SDGs will be reviewed here in the perspective of national and local administrations.

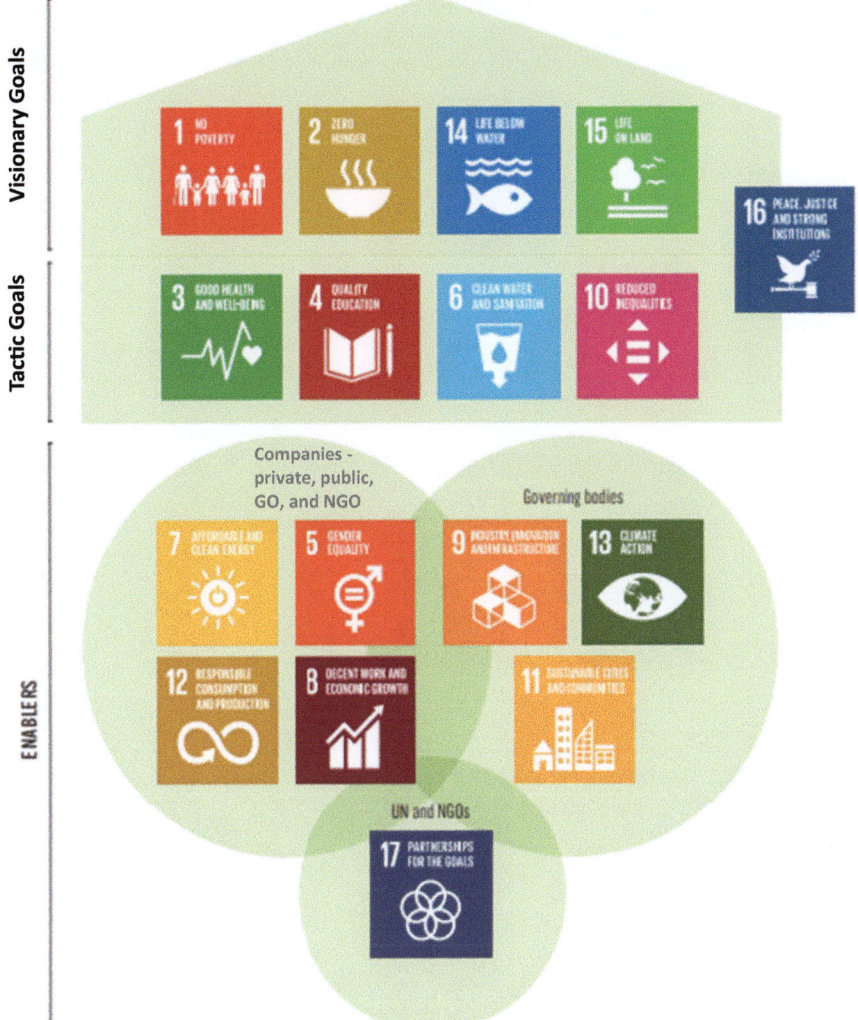

Fig. 9.1 The SDG strategy house (as in Chap. 8)

Industry, Innovation, and Infrastructure

THE SUSTAINABLE DEVELOPMENT GOALS REPORT 2023: SPECIAL EDITION- UNSTATS.UN.ORG/SDGS/REPORT/2023/

SDG#9 is on how the politicians, state, and institution supported by universities and other educational institutions build resilient infrastructure, promote inclusive and sustainable industrialization, and foster innovation, locally and with strong emphasis on the least developed countries. Three of the targets are to enhance scientific research, to upgrade the technological capabilities of industrial sectors in all countries, and in least developed countries to provide the infrastructure, the knowledge transfer, and the conditions for businesses to work and develop their strategic work of SDG#12 and 8. This really requires new ways of collaborating especially in the space of private and public sector and the educational institutions and universities.

The work that can be referred to as SDG#9 may be very strong in some countries and in some local communities, but it really is a difficult task in many countries because it first requires that the public administration rethink and innovate the way they work and interact with business and knowledge institutions.

Most of the businesses and the public administrations are hit by the financial crises, inflation, and lack of stable supply from global value chains. They are struggling to keep up and do not have the human surplus to rethink and innovate their processes and the collaborations to drive the changes. SDG#9 states that the conditions that make industries transform to sustainable business models are dependent on the framework set up by local and national regulation and actions from the public administration.

The implementation of inclusive and sustainable industrial development, resilient infrastructure, and innovation requires financing, and the UN member states have worked on creating a framework for financing the transition of the industry, fostering new infrastructure and innovation in negotiations that led to the Addis Ababa Action Agenda, that is, a global framework that seeks to align financing flows and policies with economic, social, and environmental priorities. Expanding on the previous financing for development outcomes (see https://sdgs.un.org/sites/default/files/publications/25492017ECOSOCforumonFinancingfordevelopmentfollowup.pdf).

The Addis Ababa Action Agenda highlights the need for integrated national financing frameworks to leverage the full potential of all financial flows—private and public—for sustainable development. It was developed since goals and demands were put on the least developing countries without any transfer or financing of money to be able to do so. The AAAA is trying to bridge the gap between the old, industrialized countries and the least developed countries here. Not much has happened since the declarations in 2015 and 2017, but the framework describing the needs and actions is here.

Creating sustainable transport is solved top-down from politicians on regional and national levels by implementing the infrastructure of energy-efficient and sustainable means of transport that train and shipping offers especially when shared. Companies and citizens are limited to choosing the means of transportation available. No doubt that prices are the dominating parameter here and have contributed to a shift from very efficient transport of trains and shipping to much less sustainable and inefficient transport by cars and trucks. The freight infrastructure on rails and through harbors have disappeared or is not maintained in the old, industrialized countries over the last 2–3 decades. These types of energy efficient public transport need to be reversed if we are to create sustainable transport and livable cities and this require legislation, public investments, and PPP (public-private partnership).

SDG#9 is a very important goal to impose on the public administration and on national politicians because the action and innovation should be initiated from here.

A large challenge in the transition to a Circular Economy is the creation of a recycling infrastructure of materials, and this also requires involvement and control from local public administration. In many countries, household waste handling is administrated by the municipalities, and they need to facilitate the infrastructure for a Circular Economy, also covered by this SDG.

Sustainable Cities and Communities

THE SUSTAINABLE DEVELOPMENT GOALS REPORT 2023: SPECIAL EDITION- UNSTATS.UN.ORG/SDGS/REPORT/2023/

SDG#11 is about making cities and human settlements inclusive, safe, resilient, and sustainable and importantly making cities livable again.

Approximately 1.1 billion urbans live in slums, and these are indecent living conditions, and this number is expected to rise to 2 billion in the next 30 years. Air pollution and air quality are very poor in most cities due to particle from combustion of fossil fuels. Today most cities are designed and/or occupied by cars taking up more than 50% of the space that should be made available to build decent housing, local food production, and green recreative areas that will make cities resilient towards climate change, flooding, draught, and extreme heat.

Only half of the urban citizens have access to public transportation which is one of the major problems in the unlivable cities because the cars and trucks take up space and create pollution that cause allergies and respiratory diseases and death globally.

The Global Burden of Disease (GBD) project has found that more than 5.5 million people reportedly die each year because of air pollution, making this the fourth leading cause of death worldwide. This does not include data on the diseases and mental challenges caused by noise from traffic and traffic jam, which is also severe. It is completely incomprehensible that we accept a situation like this all over the world. When implementing public transport, electrification, and regulation of individual transport would solve this in one stroke if the politicians had the courage to do so. That would require that the lobby of the oil and car industry is denied access to the politicians and their campaigning budgets. Read more on livable cities and sustainable transport in another book of the same author (Haar, The Great Transition to a Green and Circular Economy, 2024).

The planning of cities is the responsibility of local politicians and municipalities, and in most places, they hold the power to change this. The SDGs have been around for 8 years now and are due as of 2030. Some cities around the world are transforming to greener and more livable cities. A few cities in the USA have started, but cars are still at the heart of the life of American that this has way to go. The main obstrucle to implementing SDG#11 is local political will and corage. In many countries the financing of this transition can come from Public-Private-Partnerchips, but in the developing countries financing this transition is the barriere for creating fair and sustainable living conditions for people.

Climate Action

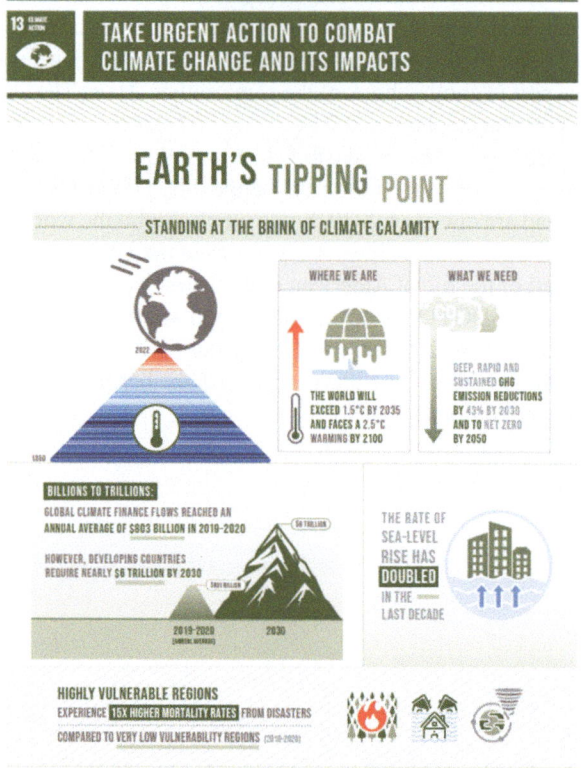

SDG#13 on climate action is not to be confused with SDG#7 on clean and affordable energy. SDG#13 is about creating the conditions on climate action at the overall level to be able to mitigate climate change and adapt to climate changes. All need to take urgent action to combat climate change and its impacts as the latest climate IPCC reports state that the surface temperature of the planet will exceed 1.5 °C by 2023 and faces a warming to 2.5 °C by 2100. Meeting the Paris Agreement targets is not possible any longer (1.5 or 2.0 °C) unless all GHG emissions are stopped by 2035. The increased temperatures have already resulted in sea level rises, and the rate of rising sea levels has doubled the last decade. The last few years, the reports on polar temperatures must mean further increase in the speed of ice melting.

Immediate action is required and finance to speed up the transition of the fossil economy to a renewable economy, and the infrastructure as well as local legislation and administration must speed up or even get started in some places. In 2019–2020 global climate financial flows reached a yearly average of $800 billion. However, the developing countries require nearly $6 trillion by 2030 to afford the transition.

Everybody is experiencing climate change, and most people talk about the need for change, and very little is happening. Energy from renewable sources and technologies is now the most affordable energy source, and the technology is available. Still climate action is lacking, and the policy makers and administrators are not rethinking and scaling the solutions at the phase necessary. It is difficult to pinpoint a few reasons for the lack of action. In Denmark it is almost impossible to have an installation of renewable energy approved by the authorities, and this could be a widespread phenomenon. It is as if those causing climate change still live to good continuing business-as-usual to drive the change.

EU has in 2019 introduced an EU Green Industrial Deal to enhance the competitiveness of Europe's clean tech industry and to accelerating the transition to climate neutrality. It includes four pillars: (1) simplified regulatory environment, (2) mobilizing private and public funding, (3) upskilling workforce, and (4) diversification of the critical raw materials (see https://commission.europa.eu/strategy-and-policy/priorities-2019-2024/european-green-deal/green-deal-industrial-plan_en).

All the three SDGs included in this chapter require change of legislation, framework conditions, and PPP (Public–Private Partnerships) with the national politicians and public administration in the chair to meet the challenges addressed. In a global perspective China must be the country that is in the front seat and speeding up on these three goals. Selected countries as Holland, Finland, and France are also examples that have changed a lot to meet the future in a sustainable way. In general, very little is happening. EU has set goal, targets, and legislation under the framework of EU Green Deal, but action is not happing at the necessary level or speed. In the Nordics, the same is the case—few cities are ramping up. Copenhagen just announced that it will miss its goals on climate neutrality in 2025, and that is only in scope 1 + 2, not even addressing scope 3. It is hard to stay optimistic on this in the perspective of the public sector and politicians.

The Nordics hold fairly low GHG emissions in scope 1 + 2, except for Denmark that has huge impacts from the intensive agricultural production of meat. When looking at scope 3, the Nordics perform poor and are on top of the list of (over) consumption, production of waste, and lack of circularity. Circularity Gap Reports on the Nordic countries have come out the latest years and the Nordic all present a circularity lower than the low global average that is around 8%. The Nordic host some of the global businesses within the *take-make-use-waste* industries and linear business models that cause overconsumption and waste production, as H&M, JYSK, Bestseller and IKEA. They deliver linear, cheap product with short lifespan creating huge amounts of wasted resources in the Nordics. Climate and environmental impacts from these business models cause impacts in scope 3 mostly in Asia. It is estimated that up to 80% of these linear businesses have their impacts in scope 3. International reports on ESG impacts often only include scope 1 + 2 data and the picture of the green Nordics is misleading due to this.

Implementing the goals presented in this chapter requiers political attention, knowledge and corage. It is very important that national and local governments show the way for businesses and at the moment it seems as if the business environment is moving the fastest. We must look at our politicians and how we elect these if the conditions for a transition are to be improved. the EU parliament has its election this year (2024) and it seams as if the will and interest for a great green transition is declining rather than increasing in the European parlament, and is shaded by the war in Ukraine and Gaza, and the movements against immigration. Overlooking the fact that climate change will cause huge migration towards Europe, and that our dependancy gas from Russian, and on imported technologies for the green transition is interconnected with the political will to transform in Europe. Adding an extra letter til ESG(Geopolitic). In US the election this year is also very formative for the transition here, as Biden has introduced the strongest and most ambitious climate regulation, globally (IRA).

The national constitutions must include paragraphs on ensuring free and open access to nature and biodiversity for all. The right to nature and biodiversity has become as important as the traditional social human rights and must be embedded into the constitutions of all countries. Humans must again become a part of nature rather than dominating nature as this will harm human habitats. Much can be learned from indigenous people and their relation and dependency to nature, as they often still live in, with and from nature and understand the necessity to care and protect nature. This is slowly being realized in the Americas but in Europe most people have lost track on the importance of being connected to nature.

Reference

Haar, G. (2024). *The Great transition to a green and circular economy.* Springer.

Chapter 10
The Sustainable Development Goals (SDGs) in a Nordic Society Perspective

Companies are part of a global economy and global value chains of products, sometimes fostering global inequality. Understanding the depth of these consequenses of the global value chains are important if a company, a state, a municipality, or an organization want to work with especially the visionary SDGs. It is important not just to fall back into a local philanthropic mindset. If an organization chooses to work with the visionary goals in the SDG Strategy House, as SDGs 1, 2, or 4, it is because it delivers a unique product or solution that fosters real change of the challenges stated in these SDGs in the full value chain and in particular in the developing countries.

The first four SDGs embrace the major inequality problems in the world, but also the increasing inequality rising in the old, industrialized world. The USA and Great Britain, but also other Western countries, are struggling with major inequality and problems of poverty of working people. In some countries, society is as greatly challenges as, for example, Africa and South America and operating here makes the visionary goals important and requires new types of responsibilites for many. The SDGs are universal, and it is important that the SDGs are not used interchangeably to solve marginal problems in developed countries, because then the solutions and the issues will become diluted and belittled.

Many of the targets under the four visionary SDGs can be addressed by major global organizations because they have the outreach and the power to impact the challenges, as poverty, hunger, health, and education. Major global players such as the UN can be frontrunners and initiators together with global companies, international industry organizations, and NGOs to create the needed changes that will really solve the issues. The UN is also a significant global procurer of goods and services and should thereby create new standards for how these kinds of challenges are solved together with local and global providers. But even the publisher of the SDGs seems to struggle in facilitating implementation of SDGs, except from publishing a lot of reports.

The SDGs address some fundamental challenges also interesting in a Nordic perspective. Many are currently looking at the Nordic welfare models and at the

Nordic societies. The Nordic economies have proven strong in times of crisis and have a high degree of public welfare and with a high distribution of wealth among the entire population. Thus, inequality in the Nordic societies is low. The GINI index is calculated between 0 and 100, where 100 is a very unequal distribution, corresponding to one person owning all the values in a country. While 0 is the point where all the values are evenly and equally distributed throughout the population. In Fig. 10.1, the GINI index of countries from 2015 is held against the GINI index from 1990 to show the development from 1990 to 2015. Countries above the dotted line have increasing GINI index from 1990 to 2015, while countries on the line have an unchanged Gini index, and countries below the line have a declining GINI index. Here it shows that the Nordic countries all have low GINI indexes below 25. Norway stands out as an interesting Nordic economy to investigate, with very large, nationalized oil reserves in the North Sea. Norway has a sharply declining GINI index from 1990 to 2015. This is mainly due to the establishment of the National Oil Fund, which is a large assets owner, also in a global perspective. The National Oil Fund with the income from oil extraction is distributing this income almost evenly among the entire population. Norway is one of the richest countries in the world per citizen, and despite its very low GINI index, one of the countries with the most

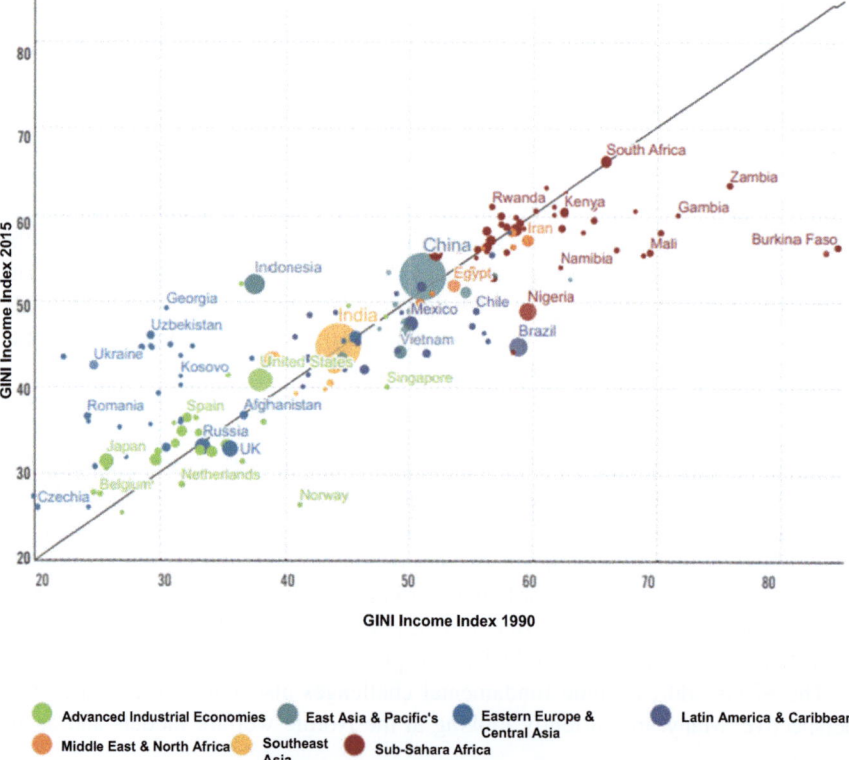

Fig. 10.1 GINI index of income in 2015 versus 1990. High GINI number equals high economic inequality. Countries above the diagonal line have an increasing GINI index and an increase in economic inequality from 2015 compared to 1990. Countries on the line have an unchanged inequality

billionaires per capita. Norway, with a population of 5.3 million people, equivalent to 0.7 per thousand of the world's population, owns 1% of all the world's wealth, mainly through this National Oil Fund. This corresponds to more than 10 times as much wealth as an average world citizen with an exceptionally low level of inequality in society. This shows that when a country nationalizes its raw materials and manages them for the benefit of its people, it is possible to be successful in distributing a great deal of wealth for the benefit of many within a very short period of time.

One could wish for the same scenario for those countries today possessing important natural resources for the Green and Circular transition, such as several African, Asian and South American countries. Unfortunately, many countries that will provide the raw materials for the future have high GINI indexes and not the democratic structures or institutions to ensure a fair spread of the exploitation of these resources, as seen in Norway.

Figure 10.1 shows that the large old, industrial countries all have rising GINI indexes, with the USA and the UK having 41 and 35, respectively, on a par with India and Russia. Despite China's strong economic growth during these 25 years (1990–2015), it has an unchanged high GINI index above 50 in 2015. The Chinese GINI is now coming down to 37 (World Bank, 2020). In the last 20 years, China has become an economic superpower and now has an economic inequality below the USA (GINI, 40; World Bank, 2020). It is important for global stability that countries holding strong economic and political power have a strong population and economic equality and a significant degree of equality. This is not the case today, where the two superpowers China and USA have GINI indexes of 37 and 40, respectively. Today the Nordic hold GINI indexes of 27–29 (World Bank, 2020) which are low rates in a European and a global perspective, with some of the most stabile economies and societies. Germany a strong economy in Europe holds a GINI of 32 (World Bank, 2019) together with France 31 (World Bank, 2020), whereas two other large economies in EU, Spain and Italy, hold GINI of 35 (World Bank, 2020), considered more unstable economies. Two other economies as the Indian and Russia are becoming powerful globally and they hold a GINI of 34 and 36, respectively (World Bank, 2021). Generally, the picture here is that the old economic power is getting more unequal, and the new economic power is becoming more equal. If this is a tendency the global economical and the political power will change over the years to come.

The most important human resource in the Nordics is trust. Only with trust the state can hold the high dominance as in the Nordics. If trust is not in place, especially trust in the state, but also trust among people, then it is not possible to spread wealth at the level done in the Nordics. In the USA and the UK, one of the largest problems is the lack of trust that has evolved over the recent decade, both the mistrust in politicians and in the states. With trust comes less corruption and the ability to tax-funded

The Nordic societies with low inequality are based on:
- Trust in the people, the state, the judiciary, and the democratic institutions (governmental and non-governmental)
- High taxation to spread wealth and create equal access to covering basic needs

- Tax-funded education, healthcare, social welfare and benefits, and infrastructure
- Safe societies with low crime rates
- Economic stability, also in times of international financial crises

and shared welfare and prosperity. Then comes a safe society, with low crime rates and economic stability. This iterator process that in the Nordic societies is illustrated in Fig. 10.2. It can be seen as a social contract that has been entered between state, citizen, and the private sector, strongly supported by pension schemes. The pension schemes in the Nordics are semiprivate as they are strongly legislated in the three-part agreements special for the labor markets here, but privately owned or member owned. This has resulted in large amount of savings kept in large pension funds under strong regulation of the state, the labor organizations, and the employer associations. These pension funds are strong and substantial institutional investors that provide investments for the private sector and create stability of the economies in the Nordics. In countries where the social contract between state, politicians, citizen, and business are broken, societies fall apart. This is what we see in the USA and UK and to some extent in European countries as Greece and some of the former Soviet states.

The Nordic countries are small countries with small populations. Apart from the dark winters and cold weather, the Nordic people share common values on their definition of a decent and free life, very different from how, for example, Americans define freedom. These shared values make economic equality possible to a larger

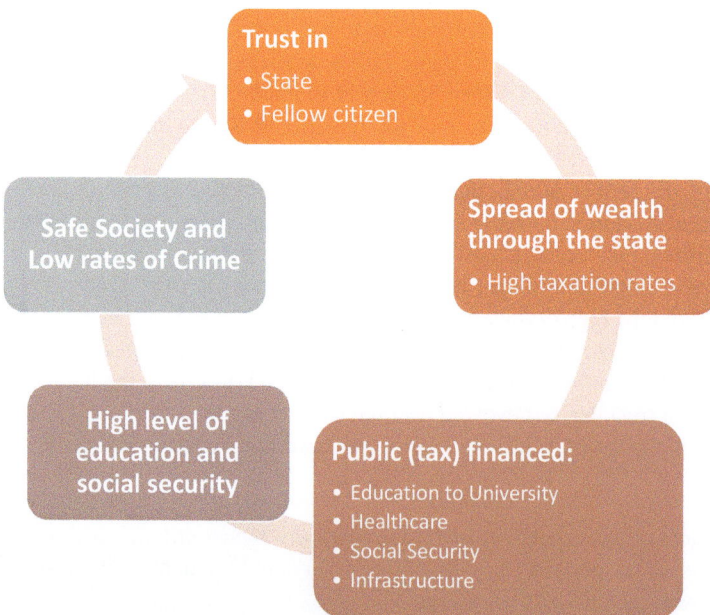

Fig. 10.2 The basics of the Nordic society model. The Nordic society models stands of a basis of shared values in the countries

extent than in many other countries, and the economics and welfare of these countries are more stable and extensive than in many economies around the world.

In this chapter follow remarks and review of selected SDGs in a Nordic perspective and may serve as inspiration for politicians and business leaders in their work with sustainability, business environment, and societies.

No Poverty

THE SUSTAINABLE DEVELOPMENT GOALS REPORT 2023: SPECIAL EDITION- UNSTATS.UN.ORG/SDGS/REPORT/2023/

Poverty is always a challenge for those affected, but it also destabilizes a society that holds large economic disparities. In countries with a high degree of poverty, a mutual disrespect arises across economic and social classes. This harms economic development and society. The socialist society model is not to promote. Even if many call the Nordic countries socialistic, they still hold a strong market economy, and in many senses the marketplace here is not as regulated as in the USA.

Extreme poverty will often lead to serious crime for survival and coverage of the basic needs of the poorest, regardless and with no respect of the people affected by the crime. In the same way, very rich people lose the touch with the challenges of common lives, even when living next door to poverty. This mutual disrespect is frightening and destabilizing as evident in many of the poorest countries. The Nordic society model is reaching global attention because of a unique cohesion, strong equality, and a high level of shared welfare, still based on market economy. Additionally, the Nordic economies and societies have proven stable, resilient to crises and safe to live in.

On a society level, it is important to cover basic needs of all people, as clearly described in the UN Human Rights Treaty (1948), and this includes abolishing poverty as the first stabilizing factor to society and for simple decency of human interaction. This is a hard task, especially in countries where corruption is widespread, because genuine politicians will be hard to put in power, even in democratic systems. South America is an example of many countries with strong socialistic movements and politicians trying to abolish poverty. Almost always spoiled by corruption on all levels in society, especially at the political level.

As described earlier here, the society contract in the Nordics is very strong and driven by a common and share understanding among citizen, politicians, and businesses of what a good society is. The political parliaments in the Nordics are multiparty democratic opposed to the two-party systems known in the Angelo-Saxon-inspired parliaments as the UK and USA. This gives a voice to many opinions and fosters cooperation in the parliament to form a government that again creates stability and cohesion.

The cohesion and social responsibility in the Nordics are expressed so well by this Danish folk singer.

"It's not the rich who are the problem. It is the existence of poor people."—Kim Larsen, Danish Folk singer

This embeds the Nordic soul and the definition of freedom. We do not accept poverty in the Nordic countries, and everyone must have the freedom to change their life situation through education and social welfare. This benefits everyone in society, also the wealthiest.

If a corporation wants to work with SDG#1, it requires a serious strategic effort in the poorest regions, either where the company operates or where it sources raw materials, goods, and services. Such an effort is challenging for a single company and must often be done in collaboration with local institutions or NGOs that have experiences with creating long-term impacts and who are able to monitor and follow up continuously. Large corporations have proven successful in showcasing how to create decent work conditions and decent lives for their workers and families in poor regions. Now many stories are out in the media on that large American corporations, as Amazon, Apple, and a.o., are exploiting local workforce in Asia, as well as in the US. Nordic cases on the same have been out on JYSK, H&M, and a.o. It is a very difficult task for the multinational corporates to operate abroad and source across continents and cultures. Some try to be responsible and manage their supply chain and others do not. With the new extended responsibilities in the extended value chains (including scope 3) and sustainability due diligence requirements legislated in the EU, operating irresponsibility will hopefully end here. Social legislation and international commitments and guidelines for corporations have been in place for at least 10 years, and now it is time to act.

Now it will be interesting to see the Nordic, global corporations leading the way to spread labor rights, common welfare, and social security all along their value chains, because they could be the best ambassadors for driving social change and abolish poverty in the least-developed countries when they source here. They stand on the shoulders of strong society models and have all the potentials of showing corporate responsibility in new ways.

Zero Hunger

THE SUSTAINABLE DEVELOPMENT GOALS REPORT 2023: SPECIAL EDITION- UNSTATS.UN.ORG/SDGS/REPORT/2023/

Hunger is not caused by a lack of food; it is caused by a lack of distribution of food, as UN-FAO statistics clearly show. In a global and regional context, there is enough food for everyone, even when wasting around 30% of all food produced. Hunger is caused by local lack of access to food due to human control, such as dictatorship, corruption, war, flight, or local climate change with droughts and floods ruining agricultural land. Hunger is man-made and may be solved even without further drag on nature. But it requires massive political and institutional effort. The war in Ukraine is challenging global food supply because both Ukraine and Russia are larger suppliers of food at a global level. Again, it is not the potential production of food; it is how humans manage the food produced.

Hunger and poverty are often a result of challenges addressed by other SDGs. Many would say that in the most remote parts of Africa, hunger is caused by a lack of food production, which is partly true. I have experienced for myself that food can be grown on very infertile soils, by the skilled farmers that Africans often are. The challenge here is that local farmers do not have access to fertilizer and irrigation or to sell their goods to an extent that can create a sustainable economy for them to kick-start the positive development of food production. This is due to poor infrastructure, poor regulation, poor trading conditions, and poor politicians.

Additionally, the breeding of drought-resistant crops has taken place, but commercial interests outside and inside Africa are working against the spread of these varieties to the regions where there is a lack of purchasing power and where the need is largest. Systems and raw materials to increase local food production are present, but due to corruption and undemocratic governance, they lack commercial incentives.

South America has partly the same challenges as Africa, except that the soils here are fertile and the potential yields are higher, even without new crop varieties. In South America, corruption is widespread and is the real cause of poverty here, not lack of access to good food production and fertile soils.

It is difficult to predict when climate change will cause food shortage, but it is certain that food production will be challenged in many places globally. Droughts and floods will create refugee flows towards safer areas due to lack of the ability to produce food. There is a need for stable economies, a change in the global distribution of economic power, and a great social responsibility in all regions and societies to counter the severe consequences of climate change. All regions have experienced areas where food production has been impossible due to draught and fires in the latest few years. It is becoming clear in Africa, Asia, Southern Europe, North America, and South America that the basis for food production is changing rapidly and human habitats are threatened in all regions. Migration of refugees over the past decade due to war and instability in the Middle East and North Africa provides a picture of what can be expected soon due to climate change.

The corona pandemic and the vulnerability of a global supply of food and goods show the need for change to create safe and responsible value chains and decent lives around the world. Here, the rich countries and the large economies have a special responsibility for the change towards a more stable and decent world.

In the Nordic region, there is an overproduction of food and wood exported out of the Nordics, despite the short growing seasons here. The soils are fertile, and until now water supply by rain has been sufficient. Climate change has also hit the Nordic regions with draught. Still reducing food waste is the most important element under the SDG#2. A third of all food produced is being wasted throughout the supply chains. This leaves a great potential in reducing waste and recovering nutrients and food. The spread of agriculture and forestry in the Nordics are very diverse throughout the region. Denmark has a very industrialized food production, and Finland has a very industrialized forestry for wood and paper production. Both countries are struggling with the eradication of wild nature over the last century, and both are on a special red list according to the new biodiversity commitments made by the EU. This is completely different in Norway, Sweden, and Iceland. These are very sparsely populated countries with a much more extensive food and wood production. It is now showing in Denmark and the oceans around Denmark that being one of the world's most intensive and industrialized agricultural countries, decades of negative impacts have harmed the environment. Today, Denmark holds less than 2% wild nature, and all oceans and freshwater systems around Denmark including the Baltic Sea are suffering from decades of overfishing and overfertilizing of the agricultural soils causing oxygen depletion in the waters. Almost no wild fish, land animals, or plants or natural ecosystems are left in Denmark, still producing huge amounts of pigs and cattle for export. In Norway, the intensive salmon industry, with sea fish farms, is harming the North Sea due to eutrophication (washing out nutrients) and the spread of medicine and disease from the cultivated salmons to the wild fish.

Just because the Nordics have been able to develop socially stable societies with a large degree of shared welfare and stable wealthy economies, it does not mean that the environmental disasters and irresponsible drag on nature have passed by here. Much of the wealth built here over the last century comes from primary production on land and in the oceans. Together with exploitation of oil and gas from the North Sea, especially in Norway and Denmark. This has all been on the cost of nature. The Nordic also hold some large corporations that build on global and irresponsible value chains and with business models promoting overconsumption, especially in the textile and fast fashion industries.

Short on Industrialized Primary Production and Food Waste

It is grotesque how industrial food production has resulted in enormous waste throughout the value chain, in an industry that has otherwise been efficient in using its resources and residues as by-products. The long supply chain with many links from farm to fork is a challenge in industrialized agriculture. It takes a very long time from a product is produced until it reaches the table. Many statistics show that the biggest food waste lies with the primary producer and the consumer. Much imported fresh food from far overseas means the food is not fresh when they are bought by the consumer. The solutions must be found throughout the value chain and the infrastructure we have created around industrial food production. This must change and is an issue in all the industrialized agriculture and in all countries consuming large amounts of industrial produced food.

Denmark being a small country with a large agricultural sector, the Danish food production is highly efficient with high, stable yields. Yet agriculture here is characterized by some paradoxes. Efficient industrial production and long processing chains causing a decoupling of consumers from primary production. Unfortunately, many Danish consumers do not demand quality or proximity to the primary producers, as seen in Southern Europe. In Denmark, price is often the most important purchase parameter, and we have created one of the world's most industrial farming countries based on very low prices and quality.

On the other hand, the Nordic foods are safe with no or little risk of contamination or unintended ingredients. Industrialized food production is characterized by a very high degree of traceability, which many industries could learn from. It is these elements that make the export of Nordic food products a success. The challenge is that these food products are often sold at world market prices, and no additional price is achieved for the safety and traceability of the production.

Now, the industrialized food production is causing large harm to nature on land and in water systems. Some would argue that efficient production makes agriculture in Denmark sustainable. Danish agriculture faces major challenges with eutrophication of oceans due to leaching of nutrients from fertilizer impacting vulnerable ecosystems on land and especially the fresh- and saltwater ecosystems. There is a need for rethinking the agricultural ecosystems, animal welfare and pesticides, as well as the management of what little wild nature remains here. As a movement towards these challenges, Denmark has a high organic production of over 10% of the total production and area. This is a reaction against highly industrial agriculture and agriculture dependent on pesticides, imports of large quantities of protein for animal feed and intensive animal husbandry. The organic product has to a large extent also become industrialized here. Many consumers do not realize that the animals they eat and get their milk, cheese, and eggs from never see the light of day and never go out in the wild. A huge divide has emerged between the rural and urban areas, and a distortion in the understanding of food production has become unhealthy. A need to rebuild the understanding of how food is produced and prepared to create a sustainable production is necessary in many countries also in the Nordics and especially in the highly industrialized Denmark.

Now it is proven out that local food production contributes to a sustainable and healthy production. When industrialized primary production is largely impacting the climate and the ecosystems due to the intensive animal husbandry, and grain production, there is a need for involvement of citizens in the production to initiate a sustainable transition and to activate the SDGs on agriculture as it interfaces strongly with SDG#3 on health. Industrial and processed food often turns out to be unhealthy food. Efficient industrial production of food is not necessarily the solution to a sustainable planet; perhaps it is the rather the challenge. Remembering that up to 70% of the world's food production is still locally produced.

Conclusive, SDG#2 zero hunger is an important global goal and is much more about respecting human rights and providing access to food locally, than about increasing overall food production. The planet can provide for 10 billion people if food production and especially food distribution is organized differently so that the huge overproduction of food that goes to waste is minimized and more importantly that the people that consume large amounts of meat as in Europe and USA change their diets to become healthy. Healthy diets and sustainable food production go hand in hand, especially if sourced locally and less industrialized. Read more about Sustainable, local food production in another book by the same author (Haar, Chap. 11: Transition to sustainable land use, Agriculture and Wild Nature, 2024).

Quality Education

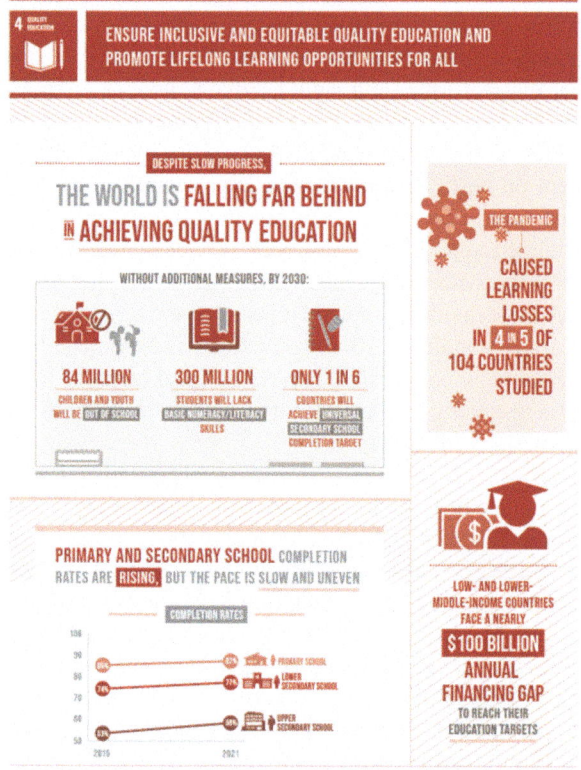

THE SUSTAINABLE DEVELOPMENT GOALS REPORT 2023: SPECIAL EDITION- UNSTATS.UN.ORG/SDGS/REPORT/2023/

Many leaders and researchers point out that education and education alone is the key to solving most of the SDGs within inequality, hunger, lack of health, and corruption, especially in developing countries. Thus, companies also have a special obligation to ensure the right educations are made available in the countries they operate. In many countries, there is no equal and free access to education for all, and thus the access to qualified labor at all levels is a challenge. Companies and people have shared values that should be addressed together with the large global and local NGOs, such as the UN and others. Access to quality education is not just a human right; it benefits people, societies, and businesses.

A high level of public education and free access to all educational levels from primary school, high schools, to vocational schools and universities are an important part of cohesion of the Nordic countries. SDG targets 4.1 to 4.7 address precisely the importance of equal access to free schools at all levels. This is offered in the Nordic countries and is probably the most important pillar in the success of the Nordics. The SDGs also address the need for equal educational for the economic success of a country. So, SDG#4 is a key to solving many of the other SDGs. In the Nordic countries, it is one of the main prerequisites for social cohesion and certainly also one of the most important prerequisites for creating sustainable development, both socially, environmentally, and economically. This is probably one of the main reasons why many political leaders are oriented towards the Nordics and the society model here. The foundation for a well-functioning society is the basic understanding of how democracy is maintained, developed, and nursed to function, is laid in the early years in primary school, and is nourished during a long education and a general high average level of education in a population. SDG#4 is where other countries and leaders can learn most from the Nordics and education build trust which is the second most important ingredient in the Nordics societies.

Decent Work and Economic Growth

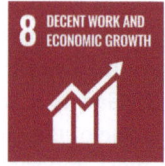

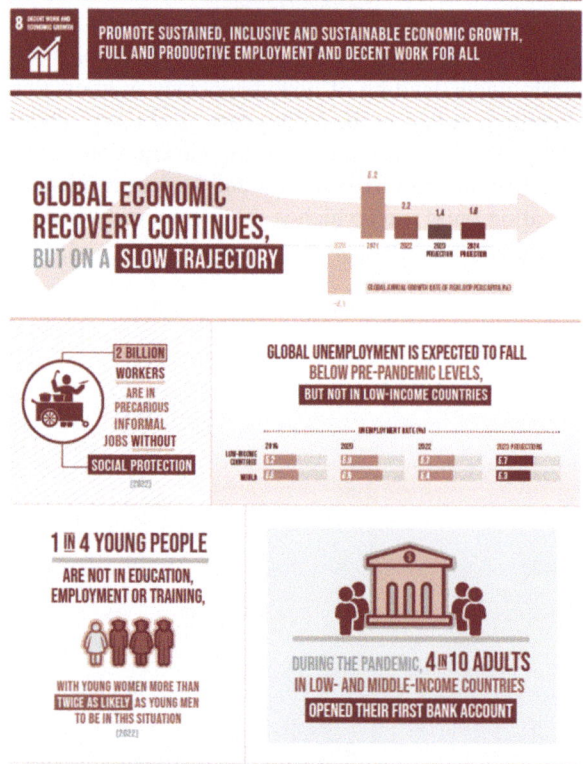

THE SUSTAINABLE DEVELOPMENT GOALS REPORT 2023: SPECIAL EDITION- UNSTATS.UN.ORG/SDGS/REPORT/2023/

In the Nordic countries, the regulation of labor conditions is strict and comprehensive, ensuring decent working conditions and wages.

In the Nordic countries, the conditions and wages are negotiated and agreed in a special collective tripartite agreement between the government, trade unions, and employers' organizations, supported by state-financed social welfare. This is a unique model and has created a common ground on negotiating labor conditions, opposite to the fight between unions and employers in many other countries. This has ensured a high degree of labor mobility and flexibility for companies to adapt the workforce to changing business conditions, as well as the social welfare of public benefits in the event of unemployment. This has also ensured uniform overall standards on working conditions a wage regulation through the labor market that has ensured good working conditions and a minimal of strikes through the last decades. Sweden has a stronger regulatory framework than the rest that are based more on 3–4 yearly negotiations. This tripartite collective model varies in the Nordic countries and were all fostered by strong labor organizations through the last century that are widely accepted by all political parties and employer-organizations. This is what Tesla (Elon Musk) is challenging by not being willing to enter agreements with the local labor organizations. He will not succeed as this is a very strong part of the Nordic market, and the unions are teaming up across countries. Tesla will not meet understanding or support by any of the tripartite as all commit to this negotiation model and the labor market is created in the Nordics.

The strong organizational power of unifying is also characteristic for the Nordics and much of the community work is organized in private organized and driven associations, whereas most of the social welfare is organized and driven by municipalities. The ability to organize and engage on community level is important in maintaining and developing the special society model here.

Working with the implementation of the SDGs around the world, inspiration can be drawn from these Nordic labor models. Nevertheless, the Nordic companies have been very progressive in outsourcing production to Asia to achieve the benefits of low wages and low environmental requirements. This will be challenged in the coming years with the extensive EU regulation on ESG, ESG due diligence, and Circular Economy. Several companies are taking back production towards Europe to gain flexibility and control over production methods and supplies, but also to ensure sustainable and transparent manufacturing. Even if the Nordic countries have created a society model with a strong spread of welfare, the countries here are challenged in the environmental part of ESG. The consumption of resources, greenhouse gas emissions, and use of nature are not sustainable.

Even the Nordic labor markets are challenged by digitalization, robotics, AI, and automation technologies replacing the grueling jobs but can also create a new type of inequality between those who manage the digital tools and those who do not. The digital tools, SoMe, and unlimited access and sharing of information on the Internet is causing mental health problems all over the world and to a large extent in the Nordics, as the Nordics has been in the forefront of digitalization.

A huge wave of especially young women and girls are experiencing mental issues. When asked they state SoMe, and the spread of private information is a significant reason for this. We haven't understood the full consequences of our online connectivity and the unregulated and free access to retrieve and share information instantly, but this will be a major problem that must be addressed and companies proving these services need to take responsibility for the consequences to larger extent. Employers must also take responsibility for the consequences of having employees online around the clock to solve work tasks. A whole new era of creating decent jobs has opened.

In the future, much of the physical work will be done by robots, and the administrative work will be done by freelancers, also called gig workers. The corona crisis has paved the way for more gig workers and new types of work agreements and conditions with the requirements to work from home for many. Less transport and much more online activity is the result of the increasing online communication. Gig work means an increasing proportion of the workforce is independent and without fixed contracts and short project-based contracts with the companies. In some societies, this share makes up to 1/3 of the total labor force, and new gig economies are developing. This new economy and way of working challenges the working rights and the welfare system because unemployment insurance and health insurance are tied to a traditional employment contract—and that people are situated close to the workplaces. This is changing with the gig economy. Social support during illness and maternity leave becomes more difficult and to negotiate on national levels, as many workers are very loosely connected to companies, to the unions, and even to the country. The strong infrastructure of the workforce is also being challenged in the Nordics these years and will challenge the welfare. The new online services and sharing economy is challenging the taxable income. Also, here new models must be developed for flexible working hours, taxation, and gig workers.

> Gig: Comes from the music industry and is slang for a single performance or a single assignment.

This new labor market with robots and freelancers requires new ways of distributing wealth in the future, if it is not to mean further centralization of power and capital away from states and to large companies. Until now, the need for human labor has been democratizing the economies. When a lot of the trade is digitized, and the manual work is automated and robotized, the economic models and the cohesion in a society are challenged. In the future, few business owners, especially in the companies based on digital business models or generating consumer data, will generate more value and cash flow than many states and thereby become very powerful. This is seen now with the larger tech-companies in the USA and in China. Global tech companies, such as Facebook, Amazon, and Alibaba, try to offer their own currencies. The control of SoMe on very few hands is also challenging the economies and societies since they do not only hold enormous amount of capital and global market share, but they also influence options and are becoming opinionator themselves. The question is whether this will be the start of an overthrow of the monetary monopoly of states, law enforcement, and other important elements of society. The balance of mutual dependencies between employee and employer known historically is changing. Very few people are extremely rich and hold extreme power, going back to the situation known from the old European and Asian Emperors, centuries ago. It must be a risk to society, democracy, the labor forces, and the international community that important society functions, as taxation, infrastructure healthcare, etc., are no longer under the control of democratic states, since they can now be carried out by a few individuals or a few large companies at a global scale. There is a need to look at new ownership models for companies and create decent jobs and responsible economic growth through ownership.

> "The world is not destroyed by those who do evil, but by those who watch the evil without doing anything."—Einstein

SDG Fund Financed by Company "Tax-Money"

Much inspiration can be collected from the Nordics. Societies based on great trust in the state and the state being responsible for education, health, and social security financed by tax income. Many leaders, political and business, can draw on this in how a transition to a fair planet is created and contributing to creating a real difference. Especially the leaders of large global, corporate businesses and private funds hold the power and the money to create significant change to a fair and sustainable planet. With the extended responsibility in the full value chain that is now imposed on companies that operate in the EU, there is a need for new ways of handling this in the sourcing countries outside the EU, but also in general committing to ESG or SDG requires new actions.

In the Nordic societies, the state is very dominant and provides all the services and thereby makes the special Nordic version of society, and this make it easy and safe as a company to operate in the Nordic if complying to local legislation and paying the local tax from the profits earned here. This is not directly replicable to other countries or communities because the people of the Nordics share the common understanding and trust in politicians and, in the state, hold a low degree of corruption, and the countries have transparent and refined legal systems. Unfortunately, this is not the case in many countries.

A progressive solution could be if the UN and the World Bank, together with international corporates, demanded governments and local authorities to comply with the SDGs, especially in the most challenged and corrupt countries. Instead of the corporates to chase linear business models, low wages followed by poor social security, cheap extraction of raw materials, and then supporting corrupt governments of the weak states. An SDG Fund financed by corporate tax payments could be established to support the local transition in the most challenge countries to responsible governance and local development of education and healthcare.

Often, companies rightly do not trust paid corporate taxes to be spend for the benefit of the population, as education and healthcare that will indirectly benefit the corporates, by providing well-trained and healthy workforce educated to demand democracy, transparency, and the absence of state corruption. Historically, the belief was that economic development also created social and democratic development, but the last decades have shown that this is not necessarily the impact caused. When not followed by demands to rulers and politicians and local business leaders on social responsibility, it will not change poverty and inequality; sometimes it will even increase this if the political and legal systems are corrupted.

With an SDG Fund financed by corporate taxes, the UN and the World Bank can launch long-term actions together with the global corporations financed by the Fund and ensure the actions to target the SDGs, such as no poverty (#1), zero hunger (#2), quality education (#4), gender equality (#5), and good healthcare systems (#3) for the populations living equally. When connected to the private investments that corporates make to create decent jobs and decent economic growth, the extended responsibility can be put into reality and drive a greater transition towards a fair and sustainable society. This requires that the UN and the World Bank can keep their hands free from corruption, themselves. This kind of development aid shifts the engagement from bilateral and region political development of state financed aid towards development through business involvement creating long-term development. Then the international companies and their brands are made more present and more vulnerable also requiring good and decent behavior. This empowerment of the global corporations creates new and more binding engagement locally for the benefit of the global community to meet the SDGs. Imagine if the oil companies and mining companies had taken this approach when they established contracts on mining and extraction of raw materials with corrupt state leaders back in the 1960–1970s. Then the world would have look different and if they had also taken the predictions on climate change seriously the world would have look different today. All this illustrates that corporations need to be much more active in global development than until now and that they hold the power to drive change inspired by the Nordics. Still it seems as if the companies are stepping up faster than many national and local ploiticians. Especially for the SDGs described in Chap. 9 there is a lack of political will and corage to change the conditions for sustainable transport, infrastructure, innovation and coordinated climate action. We must continuously support and elect the people understanding the extend and depth of this transition, and then we must change our own private way of living.

In Part III of this book, a summary of tools and overall procedures for transforming a company and preparing a sustainability road map is available. This provides an overview of the extensive and business-critical work that the company must embark on to work strategically with the SDGs.

Reference

Haar, G. (2024). Chapter 11: Transition to sustainable land use, Agriculture and Wild Nature. In G. Haar (Ed.), *The Great transition to a green and circular economy*. Springer.

Part III
The Sustainability Journey

Part III—*Describes the Sustainability Journey and Provides Tools and Methods*—for the sustainability journey and assist companies in implementing and communicating the transition to a green and circular economy. It shows how to build a sustainability roadmap for companies for development of their sustainability strategy.

Chapter 11
Sustainability Is Complex

Sustainability is a moving target as the state of the planet, the ecosystems, and climate is changing rapidly, and the science is becoming clearer. Going back to the defnition of sustainability that was put out in 1987.

> **"Sustainable development is a development that meets the needs of present generations without compromising the ability of future generations to meet their needs."— Brundtland report, UN, 1987**

The planet is now in a state where it is not able to provide for future human generations due to climate change, the lack of wild nature and biodiversity. Some decades ago, zero impact would have been enough; this is not enough any longer. Humans need to regenerate nature to ensure acceptable living conditions for the existing and future generations. The planetary boundaries are exceeded due to human activity on six out of nine parameters, and this must be regenerated so human activity is kept within the planetary boundaries again (Azote for Stockholm Resilience Centre, 2023 (2009, 2015)).

Climate change is now happing so rapidly and extensively that it is not realistic to meet the Paris Agreement scenarios of 1.5 or 2 °C unless the world cut 2030 emissions by 28% to get on a least-cost pathway for the 2 °C goal and 42% for the 1.5 °C goal (UNEP, November, 2023). As well as stop emissions completely by 2035. This recent Gap Emission Report from UNEP make frightening predictions on the scenarios of 1.5, 1.8, and 2.0 °C temperature increase, illustrated in Fig. 11.1.

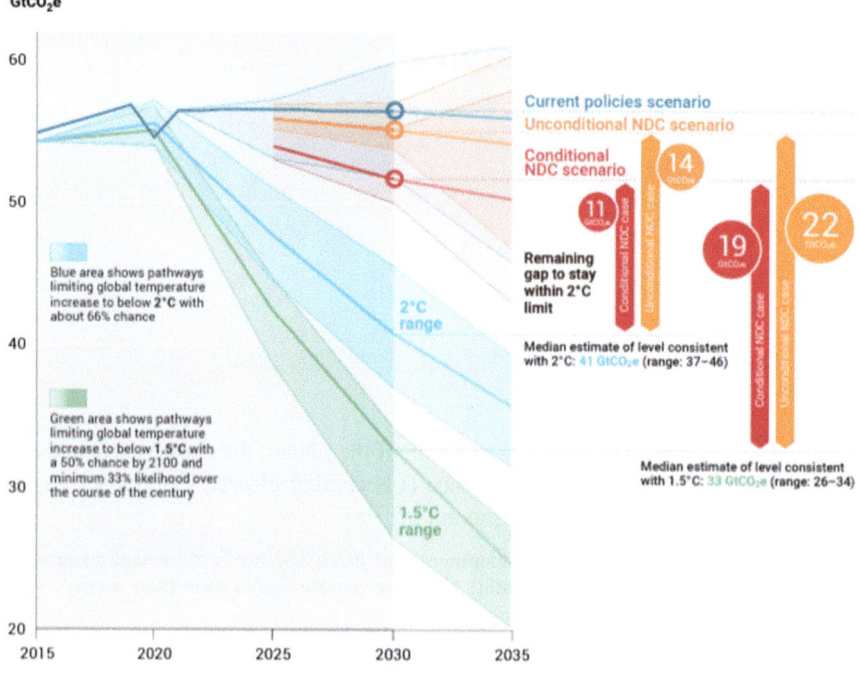

Fig. 11.1 Global GHG emission under different scenarios and the emissions gap in 2030 and 2035. The GHG emission gaps are huge and require immediate action. (Median estimate and tenth to ninetieth percentile range)

Key messages from the recent UNEP Gap Emissions Report (2023)
- The world is setting alarming emissions and temperature records, which intensify extreme weather events and other climate impacts across the globe.
- This year, until the beginning of October, 86 days were recorded with temperatures over 1.5 °C above pre-industrial levels. September was the hottest recorded month, with global average temperatures 1.8 °C above pre-industrial levels.
- Global greenhouse gas (GHG) emissions increased by 1.2 per cent from 2021 to 2022 to reach a new record of 57.4 Gt of carbon dioxide equivalent (GtCO2e).
- Similarly, GHG emissions across the G20 increased by 1.2 per cent in 2022.
- Emissions remain unequally distributed within and between countries, reflecting global patterns of inequality.

Despite the accelerating climate disasters, insufficient mitigation efforts mean the world is on track for a temperature rise far beyond agreed climate goals during this century:

- If mitigation efforts implied by current policies are continued, global warming will be limited to 3 °C above pre-industrial levels throughout this century.
- Fully implementing and continuing efforts implied by unconditional nationally determined contributions (NDCs) would put the world on track for limiting temperature rise to 2.9 °C.
- The additional achievement and continuation of conditional Nationally Determined Contributions (NDCs) would lead to temperatures not exceeding 2.5 °C above pre-industrial levels.
- In the most optimistic scenario, where all conditional NDCs and net zero pledges are met, limiting temperature rise to 2.0 °C could be achieved. However, net-zero pledges are not currently considered credible: none of the G20 countries are reducing emissions at a pace consistent with their net-zero targets.
- In the most optimistic scenario, the likelihood of limiting warming to 1.5 °C is only 14 per cent.

Source: (UNEP, November, 2023)

Climate change will negatively affect the ecosystems, and food systems are to break down within few years. The solutions are to regenerate the ecosystems and wild nature and to reestablish human living within the ecological boundaries of the planet with a population of +9 billion soon to become, as rapidly as possible. Companies and states still commit to zero impact, net zero impact, or sustainability. The fact is that the only responsible goal is regeneration at the highest speed. Regenerating ecosystems is also an important part of mitigating climate change but can **not** compensate for the stop of GHG emissions. The evolution of sustainability can be illustrated as in Fig. 11.2.

This transition from Harmful, or less impact through Zero Impact, Sustainable and Regenerative is long and requires immediate action, as well as holistic approaches. Understanding how the different sustainability impacts and actions interact is complex. Read more about the connections between climate impacts and a holistic approach towards sustainability (the Climate Nexus) in another book by the same author (Haar, The Great Transition to a Green and Circular Economy, 2024).

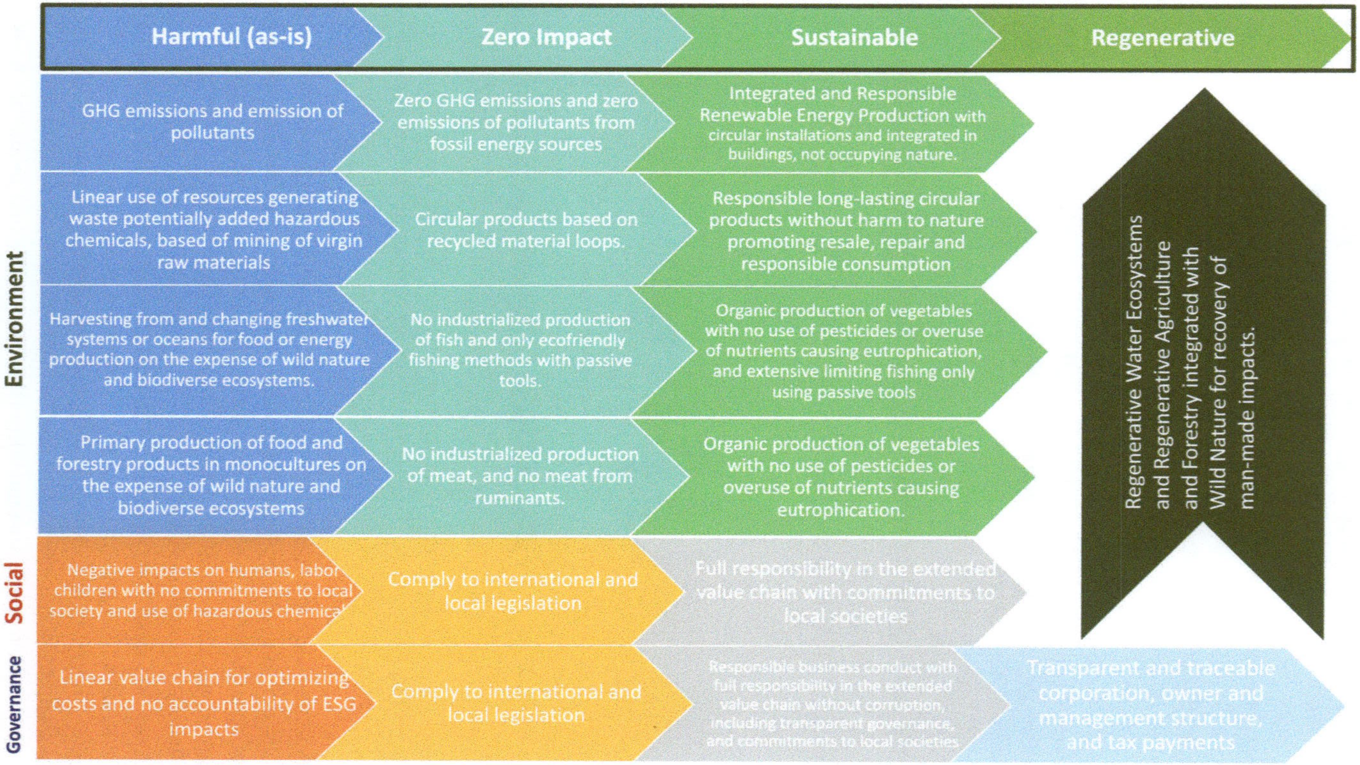

	Harmful (as-is)	Zero Impact	Sustainable	Regenerative
Environment	GHG emissions and emission of pollutants	Zero GHG emissions and zero emissions of pollutants from fossil energy sources	Integrated and Responsible Renewable Energy Production with circular installations and integrated in buildings, not occupying nature.	Regenerative Water Ecosystems and Regenerative Agriculture and Forestry integrated with Wild Nature for recovery of man-made impacts.
	Linear use of resources generating waste potentially added hazardous chemicals, based of mining of virgin raw materials	Circular products based on recycled material loops.	Responsible long-lasting circular products without harm to nature promoting resale, repair and responsible consumption	
	Harvesting from and changing freshwater systems or oceans for food or energy production on the expense of wild nature and biodiverse ecosystems.	No industrialized production of fish and only ecofriendly fishing methods with passive tools	Organic production of vegetables with no use of pesticides or overuse of nutrients causing eutrophication, and extensive limiting fishing only using passive tools	
	Primary production of food and forestry products in monocultures on the expense of wild nature and biodiverse ecosystems	No industrialized production of meat, and no meat from ruminants.	Organic production of vegetables with no use of pesticides or overuse of nutrients causing eutrophication.	
Social	Negative impacts on humans, labor children with no commitments to local society and use of hazardous chemicals	Comply to international and local legislation	Full responsibility in the extended value chain with commitments to local societies	
Governance	Linear value chain for optimizing costs and no accountability of ESG impacts	Comply to international and local legislation	Responsible business conduct with full responsibility in the extended value chain without corruption, including transparent governance, and commitments to local societies	Transparent and traceable corporation, owner and management structure, and tax payments

Fig. 11.2 From harmful to regenerative. Due to lack of action, the planet is now in the need for regeneration of the ecosystems to be able to provide for human living and to counter for the large impacts from climate change

Today the most responsible corporations provide a share of their profit to buy land or forest to regenerate ecosystems here. States also support or directly buy out agricultural land or ocean areas to allocate these to natural reserves for regeneration. The latest example is the Californian clothing company Patagonia where the owner has dedicated to whole company only to provide for regenerating nature and countering climate change through transferal of shares to a trust fund. He says that the Earth is now the shareholder. Patagonia is also seen as one the corporations in the fashion industry that has taken responsibility for the full value chain. They have also donated shares of the profit or revenue to NGO's working within natural restoration. See https://www.patagonia.com/company-history/.

Offsetting and Compensating Is Not Enough

The need for regeneration requires that using donations cannot be a part of offsetting or compensating company impacts from operations. Then the result is zero impact and not regeneration as necessary. Companies have for many years used compensation schemes or offsetting schemes to compensate for their climate emissions. The trade with carbon is a growing business but still not enough and will not make us meet the Paris Agreement nor creating regenerative ecosystems. Now is the time to do both—fully account for and remove the impacts in the full value chain and become net carbon neutral in the full value chain *and* contribute to creating regenerative ecosystems.

Human development started living in and from the forest or scrubland going thousands of years back. Then farming became the way to provide for human spread, and now humans dominate more than 70% of the land area and have cleared forests and ecosystems, as well as exploited natural resources in a way that is now threatening human existence. Now it is time to regenerate the ecosystems and change the way of living to a symbiosis with the nature and ecosystems that we are so dependent upon.

There are many tools and guidelines available for working with sustainability. For many years, the sustainability work of the large corporations has been driven by regulating the disclosure requirements of the external reporting (Non-financial Reporting Directive), together with increasing requirements from financial marketplaces as large investment banks and stock exchanges. The global audit firms have contributed to this agenda on reporting requirements, together with the UN bodies. They have been instrumental in the development of the UN Global Compact, GRI, and the tools following these initiatives. Science Based Target is one the initiatives that more and more companies are committing to and that is extending scope from GHG emissions to other ESG elements. SBTi is now being criticed by scientist of not being as science-based and not use the reconqnized methods on lifecycle assessment (LCA). The framework under the new EU Sustainable Production Regulation seems more sciencebased. The recommendation is that companies use that framework, not only because it is legislative in the EU, but also because it is thought out, requires third-party documentation, will be based on standardized LCA methods, and will create a dataecosystem to the benefit of companies, the public sector, and consumers.

Denmark was one of the first countries to legislate on company reporting on corporate social responsibility (CSR) in the Danish Financial Statements Act applicable to the largest companies. As early as in 2009 and with a continuous tightening of the rules, companies must disclose more and more on gender, responsibility, and sustainability. The result was, as expected, that companies began to report on their corporate social responsibility. Unfortunately, research showed that many of those reporting initiative did not change anything. In 2013, inspired by Danish legislation, the EU adopted the directive on reporting of corporate social responsibility applicable to the largest companies in the EU. Thus, for several years this created some awareness of what the legislators expected companies to undertake in terms of social responsibility. The reporting requirements have probably helped raise awareness of sustainability and green transition, but there is still a long way to reach the international goals set on climate mitigation, biodiversity, and wild nature. As well as meeting international agreements on the social impacts in the global value chains is still challenged by the way that the global corporates source. Even if international guidelines and agreements have been in place for years.

Sustainability has not moved properly into the strategic space of businesses or reached the top of board meetings' agendas yet. It is still considered compliance or risk management in most companies. Many leaders are now talking about sustainability and climate change, but it is still on a sidetrack of real business. Despite new disclosure requirements, international climate meetings with business representation (COP & WEF). A huge lot of scientific data on the poor state of the planet have been publicly available for decades, and still companies are not really taking the great transition to a Green and Circular Economy seriously. Leaders are like most human beings, conservative and reluctant to change, but it still seems as they have turned a deaf ear towards something that will change the market conditions and has great impact on the future business and even the existence of the company.

The green transition and sustainability are increasingly demanded by customers and needed to cover for resource scarcity. The battle for resources has become global geopolitics and the largest obstacle to business these days, just as we need to take climate change and the loss of biodiversity seriously. The access to the markets will be subject to new regulation as seen in the EU, and the green transition adds new costs and new risks to companies. Companies are also to deal with the growing global awareness of human inequality, especially when inequality is linked to food supply and natural and environmental disasters.

In a global perspective, it is important to understand that it is the few rich people that account for the major negative impacts on the planet. On climate change, it is 10% that accounts for 49% of all emissions, and 50% of the global population account for 10% of the GHG emissions, as illustrated in Fig. 11.3 (OXFAM, 2015).

This puts a strong responsibility on the 10% causing climate change, typically in the old, industrialized countries and the corporations serving here dominating the global economy, and an extended responsibility on the next 2 deciles (20%). Immediate actions implemented by the top 30% of the income would make the

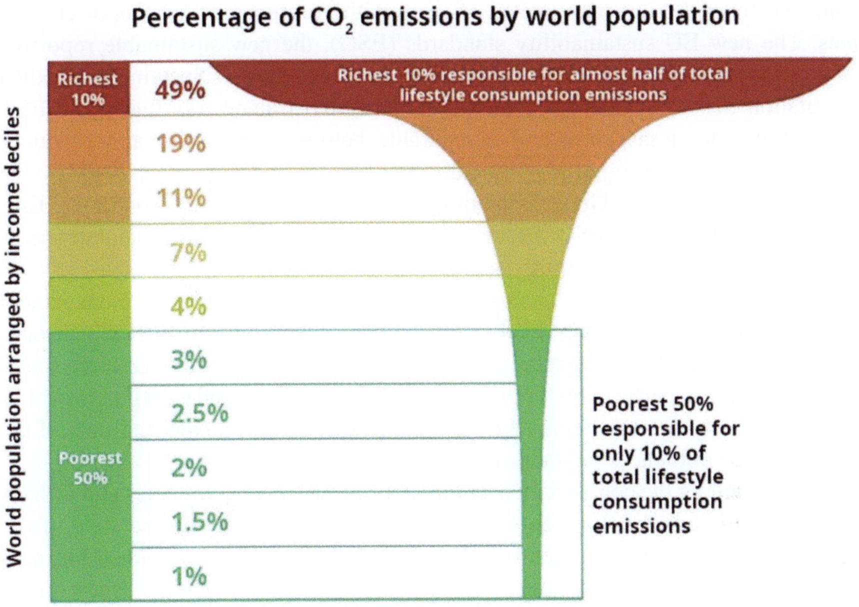

Percentage of CO₂ emissions by world population

World population arranged by income deciles		
Richest 10%	49%	Richest 10% responsible for almost half of total lifestyle consumption emissions
	19%	
	11%	
	7%	
	4%	
	3%	
	2.5%	Poorest 50% responsible for only 10% of total lifestyle consumption emissions
Poorest 50%	2%	
	1.5%	
	1%	

Fig. 11.3 Global income deciles and associated lifestyle consumption emissions. Illustrating the uneven distribution of emission impacts compared to income

necessary changes. The SDGs were developed in the light of this global inequality, not only economic and social inequality, but also on inequality within environmental impacts and rights to nature to support human lives. The SDGs solve the educational task and provide insights to the challenges and the need for change in a broader, global perspective.

The voice of Greta Thunberg about the politicians' grotesque lack of will (and knowledge) to drive change has struck many—including business leaders. Interesting is her naïve, slightly autistic rhetoric emphasizing the grotesque lack of action by the world leaders. All the data she presents are well-documented facts presented by UN, IPCC, the Environmental Agencies in EU and USA, and other international research bodies. All publicly available reports and data. At the same time, she has stumbled into a lurking and strong youth movement or a strong youth movement globally that has given the need for action wings. She has become the voice of a peer generation's powerful call to the "adults" for action.

There are still many young people who live an excessive lifestyle and overconsuming unsustainable products. But still a large proportion of very young teenagers come with a completely different awareness and fear of the future. The corona generation (those who were young teenagers then) will have a completely different starting point for understanding how the state of the planet and climate change affect them and that the world can change if we decide upon it.

Corporate social responsibility (CSR); environment, social, and governance (ESG); and triple bottom line are all "old" non-standardized concepts. These concepts cover the last many years of work with non-financial aspects of companies. The new EU sustainability standards (ESG), the new sustainable reporting standards from IFRS and GRI—all fairly aligned, and the EU Sustainable Product Regulation deliver standardized measures, KPI, and frameworks that make ESG disclosure more transparent and comparable between companies and products. Read more about the reporting standards in Chap. 5, and in another book by the same author (Haar, The Great Transition to a Green and Circular Economy, 2024). The USA is introducing ESG disclosure requirements towards financial partners as of 2024, mainly on climate impacts.

It is evident from several analysis from universities, international management consultancies, and investment banks that companies working actively with sustainability and corporate governance, with focus on impacts other than hardcore financial performance, perform better financially. Naturally, in a rapid changing world. A company is very much influenced by and dependent on society and the state of the planet, as this influences the markets strongly.

Understanding the complexity becomes a discipline in all companies and following the continuous development of scientific reports and data on the state of the planet and the climate becomes as important as understanding traditional business practices for management. Read more about competences within the green transition in Chap. 4.

References

Azote for Stockholm Resilience Centre, b. o. (2023 (2009, 2015)). *Planetary boundaries*. Persson et al and Steffen et al, Stockholm, Sweden.

Haar, G. (2024). *The Great transition to a green and circular economy*. Springer.

OXFAM. (2015). *Extreme carbon inequality*. https://www-cdn.oxfam.org/s3fs-public/file_attachments/mb-extreme-carbon-inequality-021215-en.pdf OXFAM MEDIA BRIEFING.

UNEP. (November, 2023). *Emissions gap report 2023*. UNEP.

Chapter 12
Sustainability Roadmap and SDGs

A sustainability strategy requires a detailed plan (roadmap) as any other strategy including targets, actions, and resources allocated. A sustainability roadmap also requires a baseline and decisions on when starting to measure on the various ESG/SDG parameters.

As mentioned earlier in Chap. 5, the EU sustainability disclosure regulation (SFDR and CSRD) requires both monitoring of the ESG impacts and an approach to driving the transition of companies through:

- Developing policies and procedures on the material impacts in the full value chain, including a climate transition plan. Policies must align with international agreements and commitments
- Setting targets on ESG impacts in the short, medium, and long term
- Deciding and prioritizing actions to mitigate negative ESG impacts according to a double materiality assessment of the ESG versus the financial impacts
- Allocating resources and appointing responsible in the organization
- Documenting the above in a plan approved and monitored by management—executive and nonexecutive. In reality this is a ESG-governance structure in every company

This is far-reaching requirement ensuring that this regulation is not only on yearly reporting but also on transforming the business, business models, products, and value chains. This approach is also good for companies not subject to legislation as the SME's. A ESRS standard is out for SMEs including a method for TAR (targets, action plan, resources).

Overall, this includes:

- Illustrating the full value chain based on the GHG Protocol with three scopes
- Preparing an overall double materiality assessment

- ESG reporting and monitoring
- Preparing a sustainability roadmap including targets, actions, and allocation of resources

The EU disclosure regulation also provides guidelines to implement sustainability, developing a strategy, and supported this by a strong roadmap on the how to mitigate material impacts climate, environment and social to create genuine sustainability.

The first and important task is to draw and understand the full value chain of the products and materials of a company. Thereby being able to identify and visualize the ESG/SDG impacts in this value chain. This has been proven a very strong tool, as many business leaders mainly have focused on scope 1 and 2 in their business conduct. A view on the full value chain forces management to understand, monitor, and transform the full value chain.

This chapter presents a method to build a Sustainability Roadmap based on either the ESG sustainability standards or the SDGs—or both. The EU standards are the legislative framework, and the SDGs are a good communication and education framework.

When using the SDGs as the navigator, the strategy house presented in Chap. 8 can be used to prioritize. Using the SDGs as strategic goals enables companies to develop their business and their products sustainably and meet relevant SDGs and business goals at the same time. Especially when the company chooses to work with Circular Economy in SDG#12, there is great business potential to be gained (Haar, Chap. 9: Transition to a Circular Economy, 2024d). Energy optimization and conversion to renewable energy under SDG#7 are often also profitable for companies (Haar, Chap. 8: Energy Transition, 2024c). There is more inspiration to gain from a case collection of companies in the Nordics (Haar, Nordic Case Collection on Sustainability and Circular Economy, 2024a (to be)).

> **Sustainability Roadmap (SDG Roadmap)**
> 1. **Read all the SDGs**, targets, and indicators and watch the small videos provided with the SDG educational materials on https://sdgs.un.org/goals and Part II of this book.
> 2. **Draw the company's full value chain(s)** including a full description of upstream and downstream. The full value chain is from extraction of raw materials or primary production all the way through consumption to disposal and recycling of the products. Define scope 1 + 2 + 3 according to UN climate protocol for the specific value chain. The UN GHG protocol and the three scope definitions are good inspiration to this. Read more about companies' value chains and Circular Economy in another book by the same author (Haar, Chap. 9: Transition to a Circular Economy, 2024d).

Assessing and understanding company impacts in the full value chain is necessary to meet the EU legislation and the SDGs. Legislation increasingly requires governance in the full value chain due to EU Corporate Sustainability Due Diligence Directive (CSDDD), sustainable product initiative (SPI), extended producer responsibility (EPR), and legislation on green claims. Here magnitude and materiality matters.

3. **Study the six most important SDGs in a company perspective**:

Chapter 8 in this book reviews the six most important SDGs in a company perspective, and Part II of the book walks through all the SDGs for inspiration to all types of leaders. The company must relate to each of these, but only choose 1–3 of these. Almost all companies in the old, industrialized countries must relate to SDG#12, and SDG#7 for mitigating climate impact from energy consumption and converting to renewable energy in their own buildings (scope 1 + 2).

4. **Investigate whether other SDGs are relevant** and close to the core business and the company's products. Once a company has considered the six important SDGs and found the relevant ones it is time to investigate the 11 other goals. Especially among the visionary and tactical world goals, there may be one or two goals that are close to the core business and products. SDG#16 and 17 are typically the least important in a company perspective unless the company is a large global corporate that can influence states and other institutional stakeholders. Choose only the SDGs where the company can make a significant difference at the core of its business and value chains of, through the products or services of the company. Only the goals for which business strategic goals can be set are relevant. Avoid pure off-business philanthropic initiatives. It takes away focus from the strategic work with sustainability. For example, if the company supplies water technologies, products within the fishing industry or the plastics industry, SDG#14 (life below water) must be investigated. If a company relies on or supplies agricultural products, SDG#15 (life on Land) should be addressed and so on.

5. **Conduct a double or even triple materiality assessment**. See Chap. 13 and below on the tool for the materiality assessment. Mapping materiality of company impacts (ESG impacts against financial impacts) is the first step in developing this sustainability roadmap, including the company of today (AS-IS) on the selected SDGs/ESGs and then map to incremental

subtargets and initiatives for the next 5 years (to-be) at least. This roadmap is an important part of the company's strategy.

6. **Select a total of 3–5 ESG/SDGs** as the basis for the sustainability roadmap based on the materiality assessment. The aim is to address the ESG/SDGs with the largest impact in the company value chain. For each of the selected impacts, it is necessary to:

    - Set ambitious targets on the short (1–2y), medium (3–5y), and long term(5y+) on mitigating the impacts. As a minimum, all companies must explain and prepare a plan for the material impacts identified in the materiality assessment.
    - Develop policies on mitigating the selected impacts including actions and target involving customers, employees, suppliers, and other external stakeholders. The policies must be approved and monitored by management at least on a yearly basis. One policy that is necessary for all companies to apply is a Sustainable Procurement Policy including a code-of-conduct and means of monitoring ESG in the supply chain. If the company is subject to SFD or CSRD and operating in EU, this must be supported by a sustainability due diligence. Sustainability due diligence is described in an EU directive (CSDDD), and this regulation requires companies to establish due diligence procedures to address adverse impacts of their actions on human rights and the environment along their value chains worldwide.
    - Develop and decide on actions and initiatives on the short, medium, and long term to mitigate impacts. The action plan must be detailed, and project owner (PO) and project manager (PM) must be assigned as responsible on each action.
    - Allocate resources to develop policies and drive the actions—internally and externally. Relevant resources must be assigned to the plan to ensure that implementation is feasible. Describing the actions and building the business case for implementing the actions including the internal resources and out-of-pocket costs allocated must be held against the financial wins and ESG/SDG wins gained from this.

7. **Develop a sustainability roadmap.** A template for the Sustainability Roadmap is presented later in this chapter. The format can be adapted to the company's own methods and presentations to match how the company usually communicates.

All these steps are required according to the EU regulation (CSRD and SFDR) and is a very good way to build the strategy and the Sustainability Roadmap that will enable to business to meet the future needs and the new market conditions and more importantly contribute to a fair and sustainable planet based on a Green and Circular Economy in the full value chain. See Fig. 12.1. The Sustainability Roadmap

12 Sustainability Roadmap and SDGs

SDG or ESG parameter	Targets / descriptions	SDG or ESG parameter	Status today	Baseline	Project Owner / Project Manager	Project Resources Estimate	20XX	20XX	20XX	20XX	20XX
Insert the select SDGs or ESG parameters	Describe in details SDG targets or ESG descriptions	Describe the SDG target or ESG parameter in the context of the company	Map to the existing	If available link to baseline in numbers	Project owner from Executive Management; Project Manager in the organization	Estimate the resources needed – internally and externally; Out of pocket; Time estimate					
7 (Affordable and Clean Energy)	7.2 Before 2030 the share of renewable energy in the globale energy mix must increase significantly. 7.3 Before 2030 the speed of global improvement on energy efficiensy must doubble	Has on the agenda for some time to have an energy report prepare	Has not yet had the potential uncovered	Potential not known but is estimated to be significant	Project owner: CFO NN; PM: engineer XX from production	Build a GHG baseline; Energy Optimization	Install solar panels (PV)	Climate neutral in scope 1+2			Climate neutral in scope 3
12 (Responsible Consumption and Production)	12.4 A environmental responsible handling of chemicals and waste in the full life cycle according to international agreements. 12.6 Companies, especially corporate transnational are encouraged to use sustainable practice and to integrate information on sustainability in their reporting								Implement a circular business mode		

● Targets ● Policies ● Actions

Fig. 12.1 Sustainability roadmap

is also a very good communication tool and may be the basis for the development of internal and external communication plans.

As described in Chap. 11, the planet is at state where zero impact or neutrality is not enough—now companies need to participate in regenerating ecosystems for a fair and sustainable planet. In the mapping of the value chain, the company must look at all the impacts on people and planet and profit (business) within its own walls (scope 1 + 2) as well as upstream and downstream (scope 3). Then it is possible to actively work within the company's span of control and to set demands and requirements on suppliers and customers to meet the targets set by the company. The Double Materiality Assessment (bullet 5) is relevant because the company must as a minimum choose the ESG/SDGs impacts where the company hold significant change. Responsible and sustainable companies are to transform their material impacts to the business and the ESG impacts in the full value chain illustrated by the double materiality assessment.

Understanding and Changing Company Impacts

Only by linking material ESG/SDG impacts to business impacts, as done in the Double Materiality Assessment (DMA) companies will drive change on an aggregated planetary level. The transition is not necessarily a difference measurable on a national or global level, but a significant change in the level of the company's own size and business impacts. The company must dare to risk and invest, because the sustainability is a business imperative as of now. The ESG/SDGs that the company chooses must create a long-term, measurable effect and must have strategic importance in a business perspective. Not philanthropic impacts or a little communication in the CSR report.

> **The objectives must have the same strategic weight as other strategic business decisions, meaning:**
> I. Sustainability is a top-down approach, and targets as well as actions come from executive management and are approved by the non-executive boards. Sustainability is a new business imperative, and not just a hot communication topic.
> II. Implementing sustainability changes the way a company do business—it means new business models and new products and services and new management responsibilities including scope 3.
> III. Time and money must be invested in the efforts. Sustainability must be fully aligned with other strategic business decisions.

Method to implement sustainability strategy

Figure 12.1 illustrates how manifolded a sustainability strategy is and who it is implemented.

As seen from Fig. 12.2, the sustainability Roadmap and ESG data are the drivers of implementing sustainability in a company and as illustrated the roadmap and data must continuously be updated and enriched with information from the organization, customers, and stakeholders as the transformation is happening. Learnings and new data as well as the requirements from outside will continue to influence the company and therefore sustainability and ESG will become as important as financials to manage and navigate the company into a new economy and new market conditions.

The Sustainability Roadmap is a working document that needs to be updated and approved at least on a yearly basis—it can be used as open communication internally and externally or it can stand as a working document that is the foundation of the following:

Communication plan internally and externally, supplementing the annual report, as part of the future integrated reporting that companies must submit. Many companies are these years requested by customers in their sustainability due diligence process to provide details on sustainability, and here the sustainability roadmap provides most of what it is needed together with the ESG data reporting and monitoring (non-financial data). With the communication on ESG, it is important to comply with the future regulation on Green Claims, and here companies must communicate precisely and specifically on documented achievements based on LCA (life cycle analysis). Using the words sustainable or green as a general praising of a product or a company is not allowed. This is good and it is also expected that a company communicate on targets and actions, but this itself does not make sustainability. Sustainability is based on zero or regenerative impacts on specific ESG parameters documented in the full value chain and verified by third party. It is recommended to be open and transparent on impacts and the roadmap, but to be very careful on using general terms and praising.

Education plan. With the disclosure legislation (CSRD and SFDR) is also an obligation for companies to conduct and ensure relevant education of employees to be able to implement the Sustainability Roadmap and all the actions included here. Sometimes it may be beneficial also to provide education in the full value chain - customers and vendors. See more on competences in Chap. 4.

Chapter 5 describes all the new data that companies will need to provide in the future—on company level and on product level. It is important to look beyond the annual reporting when implementing sustainability. Monitoring is necessary at the same level as financials are monitored in every company.

> The sustainability roadmap together with on-time ESG data monitoring is the strongest management tools to prepare the company for the future.

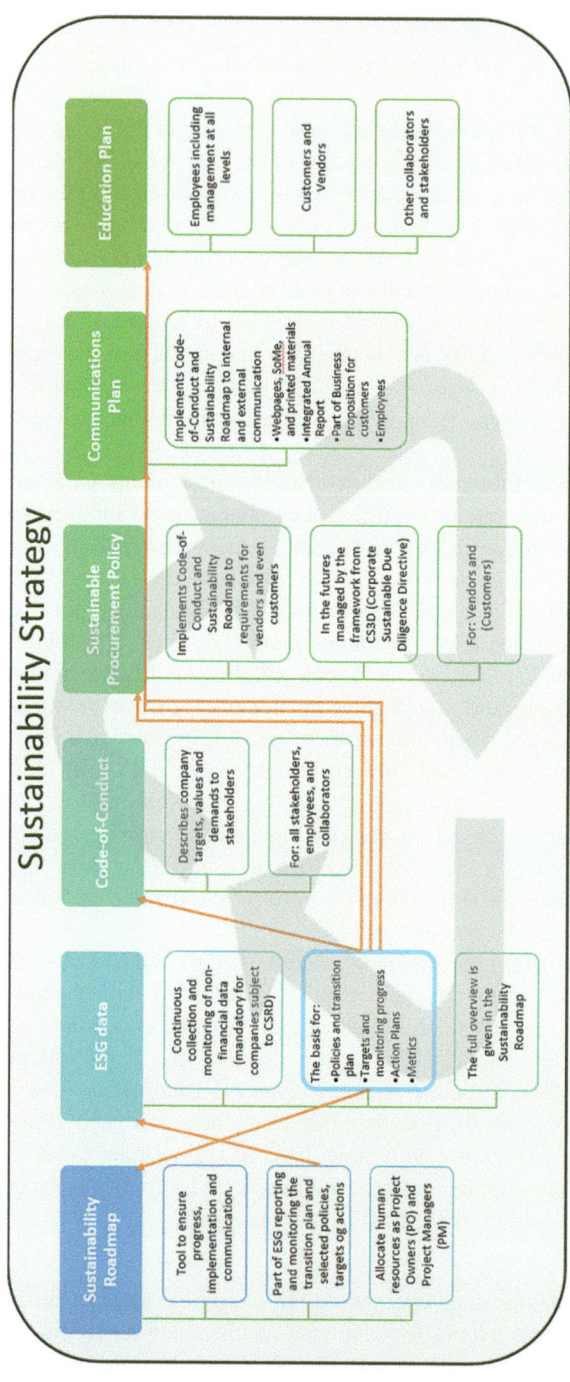

Fig. 12.2 Implementing sustainability strategy. The figure illustrates the complexity of implementing Sustainability Strategy, and that it is a continuously and ongoing process

There is an increasing tendency of greenwashing where companies map the existing activities to as many SDGs as possible and use this to communicate sustainability to customers and stakeholders. That is not the change that the planet or humanity is in the need for. It does not do anything for the company nor for the world. This is a lack of respect for the purpose and the wording of the SDGs and does not comply with the requlation in the EU. It is perfectly acceptable not to solve all the problems at once. Companies, in turn, must understand and openly communicate their impacts on the planet, climate, nature, and people and include their shortcomings and challenges in the roadmap and their communication.

Sustainability is a journey, a long journey that we are all just embarking on—governments, citizens, and businesses. It's okay to initiate the actions that make sense for the business at an affordable pace. The lack of awareness among leaders of the ESG impacts in the full value chain is a much bigger problem than not being able to solve all problems at once. The Sustainability Roadmap is a tool for documenting the implementation of sustainability. It is a roadmap that many know from other strategic processes on implementing company strategy progress. In this way, companies can create a systematic overview of how to run and prioritize their sustainability transition, leaning into the value chain descriptions and materiality analysis (MA), all to be revisited and updated at least once a year.

The Sustainability Roadmap must be adapted to the company's other strategic tools or the methods that the company normally uses for implementing strategy. Most importantly is to maintain the overview of the company's material impacts and to keep track of which areas and actions have been selected and which have been deselected. The EU regulation requires companies to explain the ESG impacts that they have not prioritized to include in their sustainability work, and to state when this is revisited. This is an educational way to force companies to take a stand on all the ESG standards continuously and create awareness of material impacts. Being able to manage material ESG impacts will be one of the most important leadership skills in the future, and here tools are necessary. This Part III presents several tools that enables executives to find their way through the new jungle of sustainability.

Many tools and advisors are on the market to support companies in this transition to support companies to not only comply but also manage the long and complex journey.

Short on International Standards for Documenting the Sustainability Journey

To support the work with sustainability, it can benefit companies to use standards and certified methods for driving the transition. Company certifications, such as ISO 14001, SA8000, ISO 26000 (SDG), and others, may be valid tools for working systematically with sustainability and in general to get started. Even for the SMEs. Although it requires time and money for assistance, certification, and audits. These tools still support the work at the tactical and operational level and will rarely get sustainability into the strategy space. This is another exercise that must be driven by management.

The Carbon Disclosure Protocol (CDP) and ISO 14064-3:2006 are tools to work specifically on climate impact and actions. Again, very comprehensive tools for SMEs to embark on. The UN Greenhouse Gas Protocol and its three scopes provide a framework for categorizing company impact at various levels, which is illustrative and easy to understand. At https://ghgprotocol.org/ tools, methods and definitions are available. The UN GHG Protocol is becoming the standard for assessing all types of sustainability impacts and is also the basis for the EU ESG standards. The future will bring international standards on ESG linked to the SDGs.

Science Based Targets initiative (SBTi) takes its basis from the GHG Protocol and is becoming a new voluntary standard that larger corporations commit to, first delivering industry-specific blueprints for climate impact and now also for biodiversity and nature. See https://sciencebasedtargets.org/. *A Corporate Accounting and Reporting Standard* published by World Resources Institute of World Business Council for Sustainable Development deliver a reporting framework for corporates to assess the climate impact according to UN GHG Protocol.

Read more about reporting and disclosure legislation in Chap. 5, or in another book by this author (Haar, Chap. 6: EU regulation to a green economy, 2024b).

References

Haar, G. (2024a (to be)). *Nordic case collection on sustainability and circular economy.* Springer.
Haar, G. (2024b). Chapter 6: EU regulation to a green economy. In G. Haar (Ed.), *The Great transition to a green and circular economy.* Springer.
Haar, G. (2024c). Chapter 8: Energy transition. In G. Haar (Ed.), *The Great transition to a green and circular econmy.* Springer.
Haar, G. (2024d). *Chapter 9: Transition to a circular economy.* Springer.

Chapter 13
Catalog of Tools and Methods

This chapter provides a selection of tools and methods to assist companies and organizations in the transition to a Green and Circular Economy and to assist them in driving the changes towards implementing sustainability that is required by legislation. The tools and methods can very well be used in combination with the Sustainability Roadmap presented in Chap. 12 of this book.

Value Chain Description

The value chain description should be supported by a graphic representation of the products and materials in full life cycle—upstream and downstream. The full value chain descriptions are the most important basis for the sustainability work. Product value chains may consist of sub-value chains—for example, food products, textile products, etc. For large corporations with numerous sub-value chains and business models, the value chains should be presented at sublevel.

Most company value chains today are linear value chains as illustrated in Fig. 13.1 (Haar, PART III: Methods and Tools for the Transition to a Circular and green Economy—Tools and Methods for companies, 2024).

In Figs. 13.2 and 13.3 are included some examples of value chains for inspiration (Haar, Chap. 9: Transition to a Circular Economy, 2024).

The goal is to transform to a circular value chain. A standard blueprint for the Circular Economy is illustrated in Fig. 13.4 (Haar, Chap. 9: Transition to a Circular Economy, 2024).

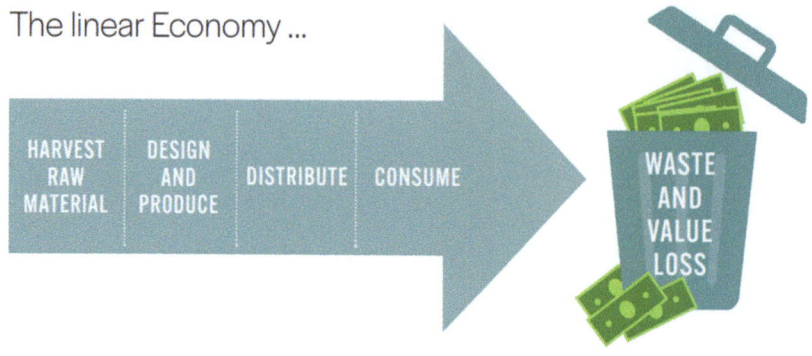

Fig. 13.1 Linear value chain

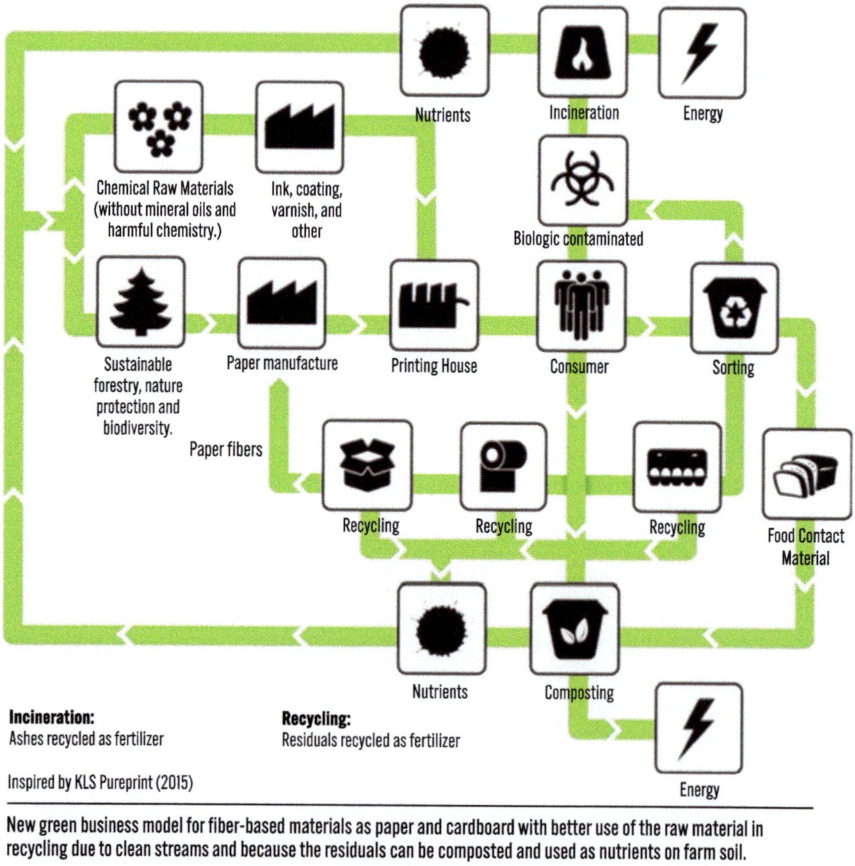

New green business model for fiber-based materials as paper and cardboard with better use of the raw material in recycling due to clean streams and because the residuals can be composted and used as nutrients on farm soil.

Fig. 13.2 Example of circular value chains for fiber materials

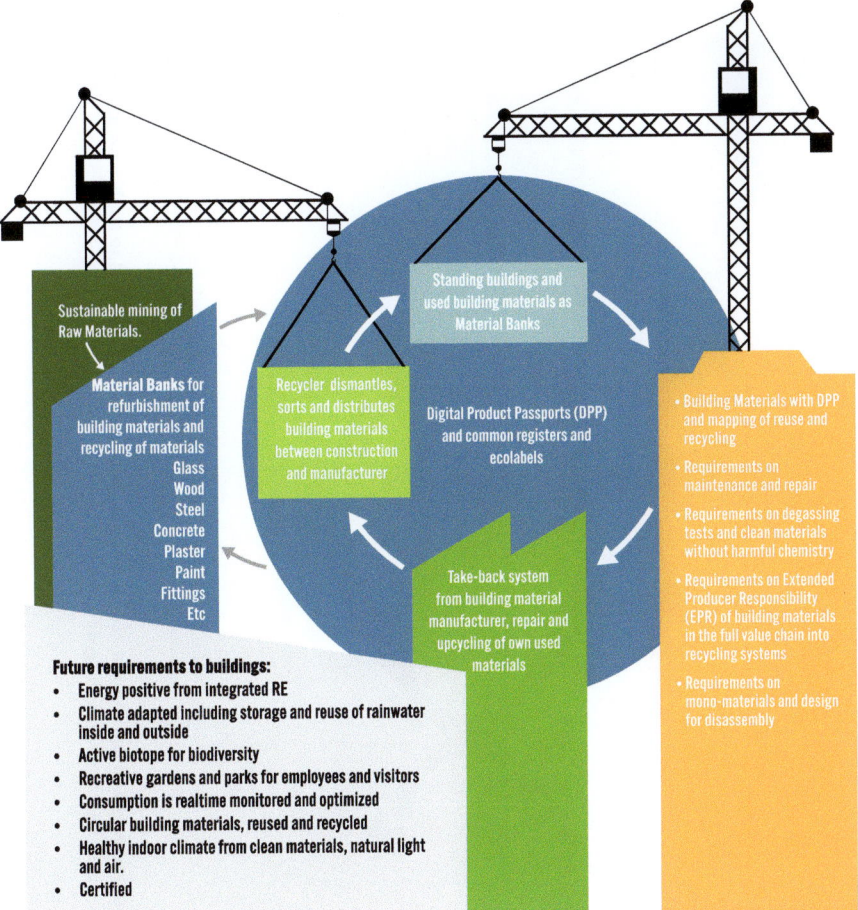

Fig. 13.3 Example of circular value chains for construction industry

Figure 13.4 illustrates the full circular value chain at product and material level to ensure reuse, recycling, and development of new business models and take-back systems. Details on transforming businesses into Circular Economy can be found in another book by the same author (Haar, Chap. 9: Transition to a Circular Economy, 2024). The overall goal with the Circular Economy is to decouple businesses and growth from the continuous exploiting of nature and virgin raw materials, as well as stopping the leak of resources as waste. Reuse and recycling in a Circular Economy provide the future raw materials from the existing. This is also important in mitigating climate impacts from the manufacturing supply chain and to minimize the impacts on wild nature and biodiversity that mining and exploiting of virgin resources has. Creating a clean and Circular Economy is key in transforming industries to sustainability and contributing to regenerating wild nature.

Standard value chains exist, for example, in the Science Based Targets framework, and will also be provided in the EU EFRAG framework later.

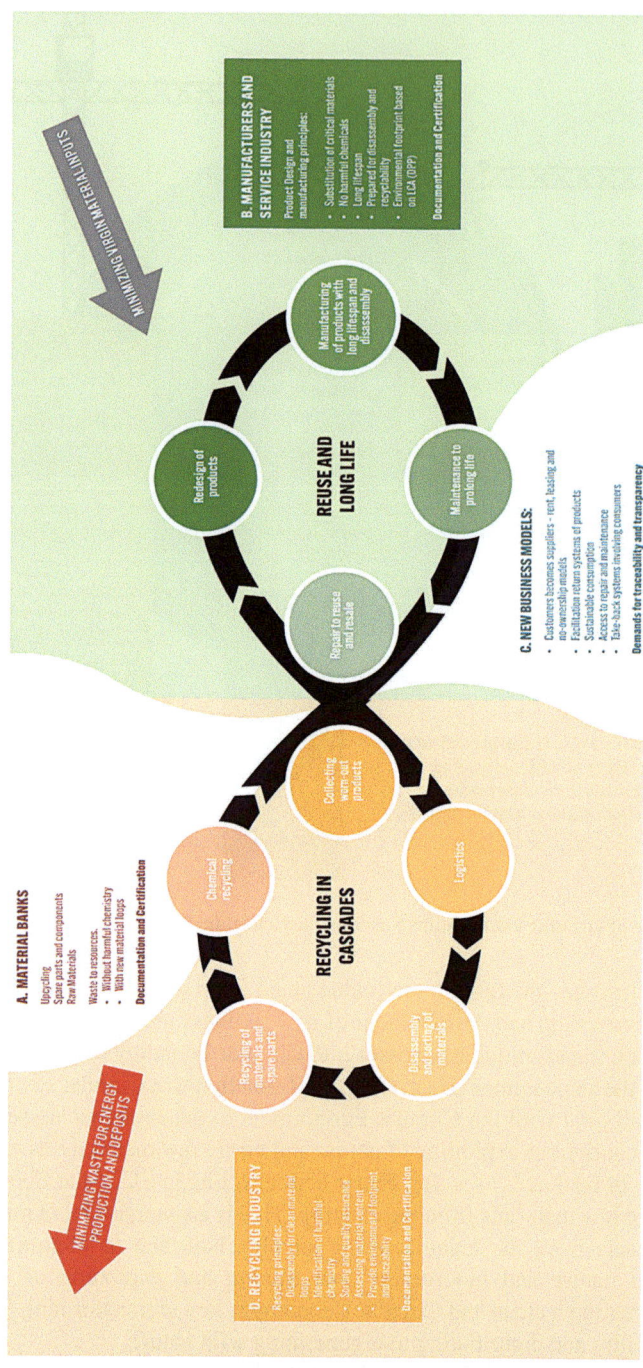

Fig. 13.4 Circular value chain. The infinity value chain illustrates a blueprint for a generic circular value chain with the goal to abolish exploiting virgin resources from nature and to stop leakage of resource as waste as illustrated with the gray and the red arrows

13 Catalog of Tools and Methods

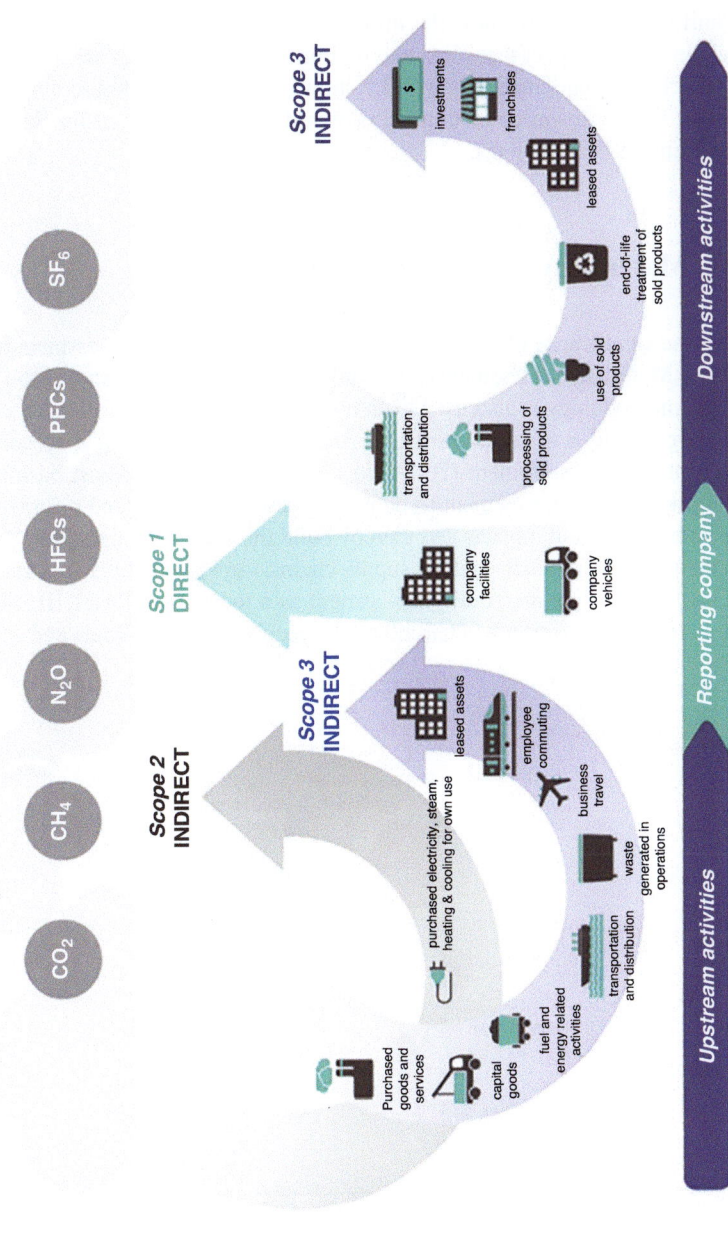

Fig. 13.5 UN GHG Protocol for companies. Detailed information on the definition of scopes can be found at https://ghgprotocol.org/

UN GHG Protocol for Companies

As mentioned in Chap. 1, the UN GHG Protocol defining the three scopes are setting a uniform standard for how the impacts of companies are assessed in the full value chain for all impact measures and definitions on the scopes are also stated in the EU disclosure regulation. Figure 13.5 shows the UN GHG Protocol for companies (WRI, World Resource Institute, and World Business Council for Sustainable Development (WBCSD), n.d.).

Double Materiality Assessment (DMA)

To assess the significance of company ESG impacts a materiality assessment must be conducted. Materiality assessment is a term inspired by audit principles to identify risks. This methodology is also used to assess ESG impacts in the full value chain to identify targets and actions. In the EU regulation, the requirement is to conduct a Double Materiality Assessment (DMA) to hold ESG impacts against business and financial impacts for management to be able to assess, evaluate, and prioritize ESG and business countering impacts and thereby build the strategy to transform the business models and the products and the supply chains—upstream and downstream.

In Fig. 13.6, a triple materiality assessment as a tool (Haar, PART III: Methods and Tools for the Transition to a Circular and green Economy—Tools and Methods

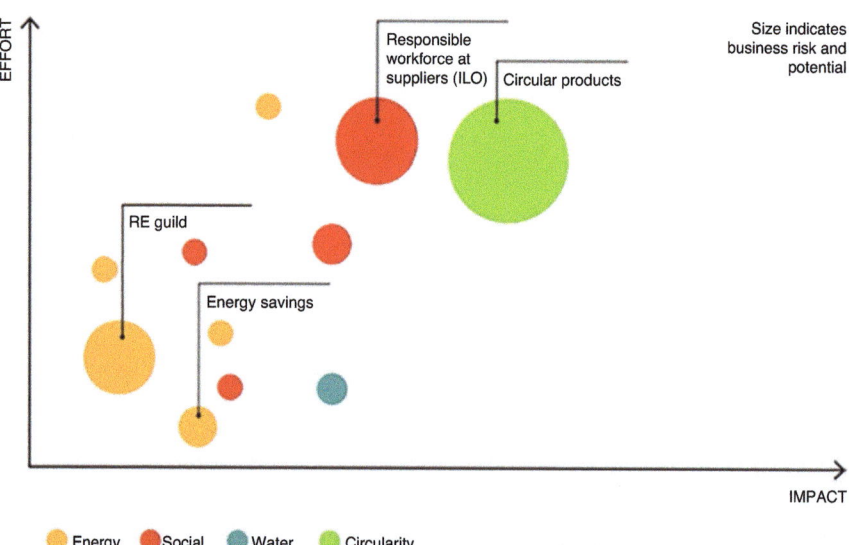

Fig. 13.6 Materiality assessment. Materiality assessment with three dimensions—impact is held against effort on the axes. The size of circles presents business risks/potentials, and the colors indicate types of impacts

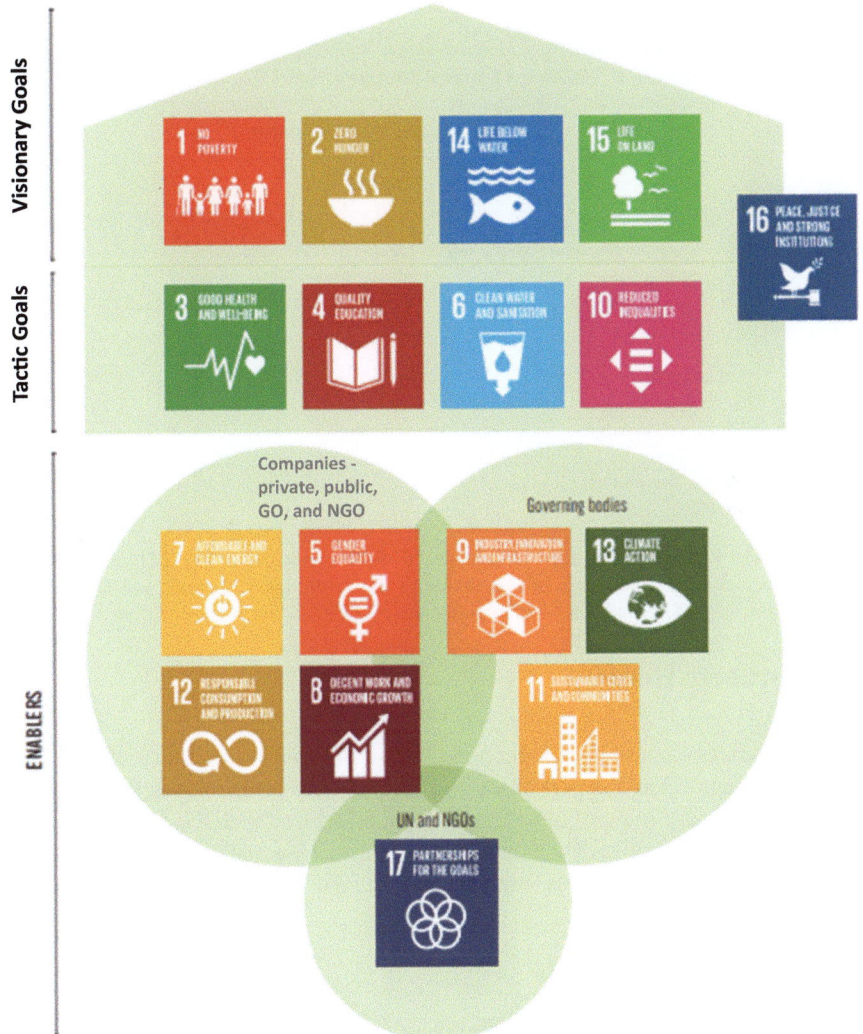

Fig. 13.7 SDG strategy house. SDG strategy house to assist companies, organizations, and municipalities to prioritize the Sustainable Development Goals

for companies, 2024). This also includes the efforts and resource allocation to implement the actions on the different impact and business measures.

The materiality assessment is the tool to prioritize corporate sustainability initiatives. Impacts are measured in relation to effort and business impact as potential risks and business opportunities for the company. A business risk on not transforming negative ESG impacts often also offers a business opportunity. Originally the materiality assessment was suggested by the GRI as a tool for stakeholder analysis to identify how stakeholders assessed the ESG parameters of the company, and then it was more of a communication tools. Now it has been transformed into a strategic business assessment tool against ESG impacts.

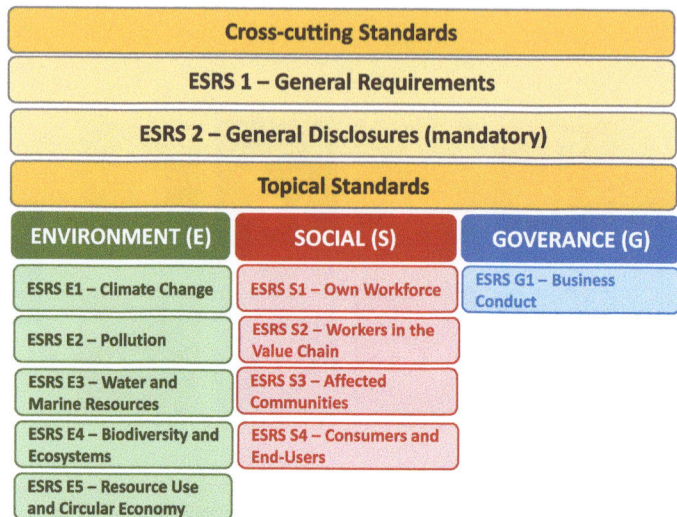

Fig. 13.8 EU ESG sustainability standards (EFRAG guidelines). The EU Sustainability Standard includes the parameters of sustainability on Environment, Social, and Governance impacts defined by ESRS and described in guidelines by EFRAG. See www.efrag.org

SDG Strategy House

The SDG strategy house is explained in Chap. 8 in this book. It also included here in this catalogue since it is part of a set of tools to help companies with the SDGs. The Strategy House is a tool to understand and prioritize the SDGs in a business or organizational perspective. Then the company can choose SDGs based on the material ESG impacts and business impacts, as ESG and SDG are aligned topics. See Figs. 13.7 and 8.1.

ESG Sustainability Standards

Figure 13.8 illustrates the EU sustainability standards (European Commission, 2024) and in details described by EFRAG (EFRAG, 2024). Here is also included the EU sustainability standards highlighting the ten parameters where companies are expected to assess their impact and then monitor, report, and drive change within the most material impacts, including the cross-cutting standards on how to organize monitor and implement the ESRS legislation. It is in these general requirements and general disclosure that the mandatory methodology is described.

The EU sustainability (ESG) standards are described shortly in Chap. 5 and in more details in another book by the same author (Haar, Chap. 6: EU regulation to a green economy, 2024). EFRAG is continuously publishing updates on the 12 guidelines that describe the reporting and monitoring requirements of the ESRS legislation.

Fig. 13.9 Minimizing climate impact in scope 1 + 2. This process is to implement energy optimization in scope 1 + 2 in companies and other organizations

Minimizing Climate Impacts in Scope 1 + 2 of Companies and Organizations

The tool illustrated in Fig. 13.9 is a tested and well-defined process to minimize GHG impacts in scope 1 + 2 by optimizing energy usage, electrification, and installation of renewable energy (Haar, PART III: Methods and Tools for the Transition to a Circular and green Economy—Tools and Methods for companies, 2024).

The process is described in details and cases are shared in another book by the same author (Haar, The Great Transition to a Green and Circular Economy, 2024).

Fig. 13.10 Transforming into a circular business model (scope 3). This process describes a proven method to transform into a circular business models minimizing ESG impacts in scope 3

Transforming the Business Model into a Circular Business Model (Scope 3)

The tool illustrated in Fig. 13.10 is a tested and well-defined process to transform business models into circular business models thereby minimizing impacts in scope 3 to save GHG, resources, and other environmental and social impacts (Haar, PART III: Methods and Tools for the Transition to a Circular and green Economy—Tools and Methods for companies, 2024).

In another book by the same author, the process is described in details, and cases are shared (Haar, The Great Transition to a Green and Circular Economy, 2024).

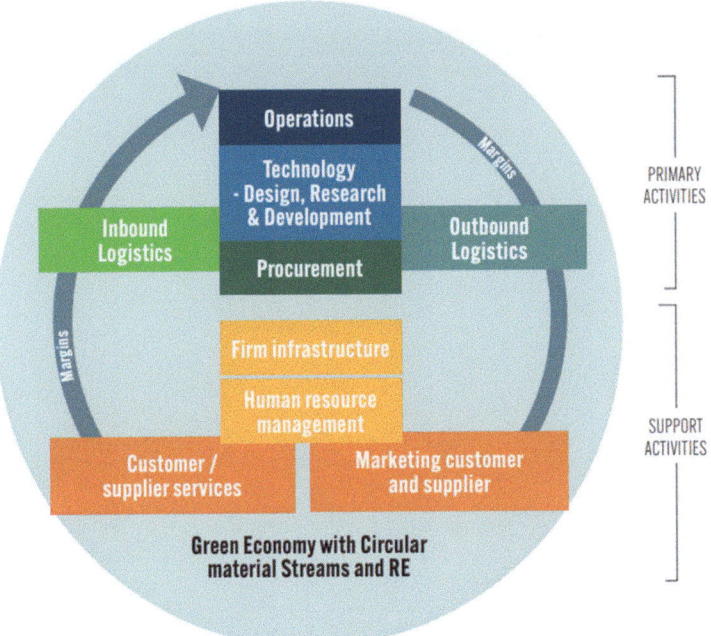

Fig. 13.11 **Circular organization model.** Reorganizing activities and support functions in companies are necessary in a company operating in a Circular Economy

Circular Organization Model

The circular organization model included in Fig. 13.11 is explained in Chap. 6 in this book and illustrated in Fig. 6.4. This is a model to understand the organization of primary and secondary activities in a company in the Circular Economy. Companies need to organize themselves differently when the economy and the market conditions changes. The model is inspired by M. Porter's value chain and is a new way of organizing the company.

All these tools are presented as a catalog that can support the implementation of the Sustainability Roadmap in companies and organizations at a strategic level, as well as at the operational and hands-on level. Detailed description of the tools in this catalogue are described in details in other chapters of this book or in another book by the same author: The Great Trasition to a Green and Circular Economy. Haar. 2024. SpringerNature.

References

Haar, G. (2024). *Chapter 9: Transition to a circular economy*. Springer.
(WRI), World Resource Institute & World Business Counsil for Sustainable Development (WBCSD). (n.d.). *GHG Protocal for companies*. Retrieved from GHG Protocol: https://ghg-protocol.org/

Chapter 14
Epilog: It Is Just a Human Choice

What If…

… cars absorbed CO_2 rather than emitting it.

… houses, buildings, and constructions produced energy rather than using it.

… all raw materials were recycled infinitely, and we never had to mine from nature again. Then nature would start regenerating and flourish rapidly.

… we all lived on healthy diets and not overconsumed sugar, fat, and meat and then regenerated nature and ecosystems.

… we shared instead of owned.

… we left the oceans without leaching nutrients from agriculture or fish farms causing eutrophication (leaching of nitrogen and phosphorous), as well as oxygen depletion and death of fish and ecosystems like mangroves and coral reefs.

… we did not overfish or pollute the oceans and freshwater systems with chemicals, plastics, and other waste products. Then the oceans again could become the largest and most important ecosystem and food system and a sink of CO_2 rather than emitting CO_2.

… we used the existing infrastructure for shared transportation and energy production and stopped constructing more roads.

… we reused all the existing building materials that has been sitting in houses for decades and centuries, and retrofitted, renovated, and reloved all these good materials that nature gave us. Rather than continuously mining from nature and generating the largest amount of waste from one single industry.

… we made cities livable for people instead of for cars and used the space for recreative nature, biodiversity, and local food production providing mental health for the people.

… we took the convenient speed-train to all cities on the continent rather than inconveniently airplanes and saved the hassle and the climate.

... we transported all cargo and goods by ships and boats and intelligently distributed from harbor storage sites to people instead of using unhealthy trucks that emit, pollute, make noise, and take up space in and around cities.

... we paid for what we took, also from nature, climate, and raw materials. It just like in the supermarket—if it was all free it would soon be finished.

... forests were wild nature and healthy ecosystems rather than production plants.

... people lived in peace and harmony (okay, let's not get unrealistic).

None of these suggestions are impossible; all the science, knowledge and technology is available to do so. It is not the poorest 60% that pushes the planet out of its boundaries; it is the riches 10%. Economy, the planet, and humanity will gain if we change our ways of living.

It is just a human choice!

WE decide if we want to change, not nature, not the climate, nor our neighbor. We can all change the way we live and make the planet fair, livable, and sustainable for all. It is just a choice, and there is only **one reason** for not making the right choice....

Let's start now!

MIX
Papier aus verantwortungsvollen Quellen
Paper from responsible sources
FSC® C105338

If you have any concerns about our products,
you can contact us on
ProductSafety@springernature.com

In case Publisher is established outside the EU,
the EU authorized representative is:
**Springer Nature Customer Service Center GmbH
Europaplatz 3, 69115 Heidelberg, Germany**

Printed by Libri Plureos GmbH
in Hamburg, Germany